Signals

Second Edition

D1101683

F R Connor

Ph D, M Sc, B Sc (Eng) Hons, ACGI,
C Eng, MIEE, MIERE, M Inst P

Edward Arnold

A division of Hodder & Stoughton

LONDON   NEW YORK   MELBOURNE   AUCKLAND

© F. R. Connor 1982

First published in Great Britain 1972
Reprinted 1975, 1978, 1979
Second Edition 1982
Reprinted 1985, 1988

Distributed in the USA by Routledge, Chapman and Hall, Inc.
29 West 35th Street, New York, NY 10001

British Library Cataloguing in Publication Data

Connor, F.R.
  Signals.—2nd ed.—(Introductory topics in
  electronics and telecommunications)
  1. Signal theory (Telecommunication)
  2. Telecommunication
  I. Title   II. Series
  623.89'4   TK5102.5

ISBN 0-7131-3458-2

Typeset by Macmillan Ltd, Banglalore
Printed and bound in Great Britain for Edward Arnold, the
educational, academic and medical publishing division of Hodder
and Stoughton Limited, 41 Bedford Square, London WC1B 3DQ
by Athenaeum Press Ltd, Newcastle upon Tyne

Preface

In this new edition, important parts of the text have been revised or extended, especially in the areas of signal processing and information theory. In Chapter 2 on signal analysis, details of the discrete Fourier transform (DFT) and the fast Fourier transform (FFT) are introduced and Chapter 3 now concludes with an introduction on the important concept of convolution for determining network response. Chapter 5 which has been renamed Signal processing includes recent developments in speech and image processing and an insight into the areas of television and picturephone processing. The final chapter on information theory contains some basic ideas on coding theory and has been extended to cover the Bose–Chaudhuri–Hocquenhem (BCH) and convolutional codes. The main text has been enlarged by the use of various appendices to cover such new topics as the Cooley–Tukey algorithm, Walsh functions, and the rate-distortion function. It is intended for the reader who seeks a deeper knowledge of the subject and it can be supplemented by means of the large set of references provided for further reading. As in the first edition, numerous worked examples are included together with a set of representative problems with answers for the interested reader.

The aim of the book is the same as in the first edition, though it should be pointed out that Higher National Certificates and Higher National Diplomas are being replaced by Higher Certificates and Higher Diplomas of the Technician Education Council.

In conclusion, the author would like to express his gratitude to those of his readers who so kindly sent in some comments and corrections for the first edition.

1982   FRC

Preface to the first edition

This is an introductory book on the important topic of *Signals*. Electrical signals in various forms are used extensively in the fields of electronics and telecommunications, and the book endeavours to present the basic ideas in a concise and coherent manner by bringing together closely related subject matter all under one heading. Moreover, to assist in the assimilation of these basic ideas, many worked examples from past examination papers have been provided to illustrate clearly the application of the fundamental theory. The book is the first of six dealing with the fundamental topics of telecommunications and electronics.

The early chapters of the book are devoted to an analysis of the various types of signals and a study of their particular characteristics. Subsequent chapters deal with the transmission of signals and the signal techniques employed in various applications. The book ends with an introduction to the important subject of information theory, which deals with the general problem of the transmission of information in any communication system.

The book will be found useful by students preparing for London University examinations, degrees of the Council of National Academic Awards, examinations of the Council of Engineering Institutions, and for other qualifications such as Higher National Certificates, Higher National Diplomas, and certain examinations of the City and Guilds of London Institute. It will also be useful to practising engineers in industry who require a ready source of basic knowledge to help them in their applied work.

1972   FRC

Acknowledgements

The author wishes to thank the Senate of the University of London and the Council of Engineering Institutions for permission to include questions from past examination papers. The solutions provided are his own and he accepts full responsibility for them.

The author would also like to thank the publishers for various useful suggestions and will be grateful to his readers for drawing his attention to any errors which may have occurred.

Contents

Symbols

c	any number
f	frequency
f_c	cut-off frequency
f_r	repetition frequency
$f(t)$	any function of time
$f(nT)$	n^{th} discrete time sample of a signal
$h(t)$	impulse response of a network
i	instantaneous current
k	any number
m	any number
n	any number
p	probability of a digit
q	charge
s	complex variable
v_g	group velocity
v_{ph}	phase velocity
A	peak amplitude
C	capacitance
	communication capacity
C_n	n^{th} Fourier series coefficient
$\mathscr{F}[f(t)]$	Fourier transform of $f(t)$
$F(\omega)$	Fourier transform function
$F(m\Omega)$	m^{th} discrete frequency component
$F(z)$	z-transform function
H	average information (entropy)
$[H]$	any matrix H
H_c	conditional entropy
H_{max}	maximum value of information H
$H(\omega)$	transfer function of a network
$H(X)$	average information of message X
$H'(X)$	average information rate of message X
$H(X\mid Y)$	average conditional information of message X transmitted given message Y is received
I	self-information
$I(X;Y)$	mutual information between message X transmitted and message Y received
L	inductance
$\mathscr{L}[f(t)]$	Laplace transform of $f(t)$
$\mathscr{L}^{-1}[F(s)]$	inverse Laplace transform of $F(s)$
N	any number
	average noise power
P_{av}	average power

P_i	probability of ith symbol
$P(x)$	probability density function of variable x
$P(x,y)$	joint probability density function of variables x and y
R	information rate
	resistance
$R(D)$	rate-distortion function for fidelity criterion D
T	periodic time
	transmission time
V_i	input voltage
V_o	output voltage
W	energy
	highest frequency component
$[W]$	any matrix W
α	any constant
β	any constant
	phase shift per unit length
$\delta(t)$	Dirac delta function (impulse function)
λ	wavelength
σ	any positive number
	rms noise power
τ	duration of time
$\phi(\omega)$	phase angle
ω	angular frequency
ω_s	angular sampling frequency

Abbreviations

C.E.I.   Council of Engineering Institutions examination in Communication Engineering, Part 2

U.L.   University of London, B Sc(Eng) examination in Telecommunication, Part 3

1
Introduction

The communication engineer is concerned with the transmission and reception of signals. A signal is an electrical voltage or current which varies with time and is used to carry messages or information from one point to another. A message is usually in the form of words or coded symbols and the amount of information it contains is of great importance in communications.

In practice, it is more convenient to handle information by converting it into a signal. The signal is then transmitted over a communication system to the receiving end, where it is transformed back to the original information or message. The schematic arrangement of a typical communication system is shown in Fig. 1.1.

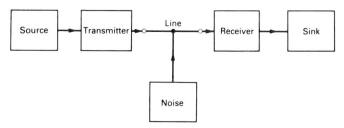

Fig. 1.1 Typical communication system

The source generates the message signal which is processed by the transmitter and sent along a transmission line. At the receiving end, the message is extracted by the receiver and sent to its final destination (sink). During transmission and reception, noise is picked up from various sources and these can be represented as a single noise source.

1.1 Types of signals

There are two main types of signals – the *analogue* signal which varies continuously with time, and the *digital* signal which is discontinuous with time.

Analogue signals usually represent the variation of a physical quantity, e.g. a sound wave, and are either single sine waves or a combination of them.

The digital signal consists basically of *pulses* which occur at discrete intervals of time. The pulses may occur singly with a definite periodicity or in groups, in the form of a code, as in telegraphy.

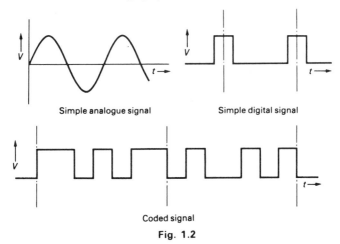

Simple analogue signal   Simple digital signal

Coded signal

Fig. 1.2

1.2   Examples of signals

Typical signals are those used in telegraphy,[1] telephony,[2] radio communication,[3] television,[4] and radar.[5]

Telegraph signal
A message consisting of a set of words may be transmitted by ascribing to each letter a certain coded signal. This is the basis of telegraphy and it is usually achieved by means of an electromechanical machine, somewhat like a typewriter and known as a teleprinter.

The letter R used in the teleprinter code has the form shown in Fig. 1.3.

This signal has a pulse waveform with positive and negative values called *mark* and *space* respectively. The rate of transmitting the pulses is called the signalling speed and is measured in 'bauds'. The baud is defined as the number of pulses sent per second and the usual speed used in machine telegraphy is about 50 to 75 bauds.

It will be shown later that such a pulse waveform consists of a range of frequencies called the *bandwidth* of the signal. For a 50 baud speed, this is taken as 120 Hz per message for practical reasons and is known as a *telegraph channel*.

Example 1.1
Discuss the relationship between bandwidth and signalling speed in a telegraph system.

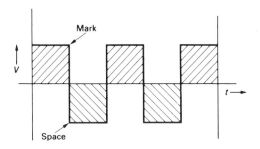

Fig. 1.3 Letter R

Solution

Consider a simple system in which the basic signal consists of a series of positive and negative pulses as shown in Fig. 1.3. The speed of signalling is defined as the number of elementary pulses sent per second and is therefore reciprocally related to pulse duration; since the narrower the pulse, the greater is the number that can be sent per second, i.e.

$$\text{signalling speed} = \frac{1}{\text{duration of elementary pulse}}$$

For example, for a signalling speed of 50 bauds, we have

$$\text{duration of elementary pulse} = 1/50 = 20 \text{ ms}$$

Now, pulse duration or time is itself *inversely* proportional to frequency and so the signalling speed is *directly* proportional to frequency. Hence, the higher the signalling speed, the greater is the frequency involved. In other words, the bandwidth of frequencies used is proportional to signalling speed.

Telephone signal

A telephone conversation consists of speech sounds involving vowels and consonants. The speech sounds produce audio waves which cause a diaphragm to vibrate in the telephone mouthpiece, thereby producing an electrical signal. The speech sounds fluctuate considerably in form and so the telephone signal consists of a complex combination of audio-frequency sine waves. The signal obtained for uttering the vowel U is shown in Fig. 1.4.

Most of the energy in speech signals is found to be contained in the lower frequencies; for intelligible speech, it is sufficient to consider those from about 300 to 3400 Hz. An overall bandwidth of 4 kHz is therefore used for each telephone message and is referred to as a telephone channel.

Alternatively, such a telephone channel may be used to carry several telegraph messages, each with a 120 Hz bandwidth. As many as 24 telegraph channels can be accommodated in one telephone channel.

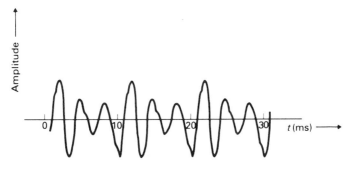

Fig. 1.4   Vowel U

Radio signal

A radio signal is generated by an oscillator and consists of a radio-frequency sine wave which is called the *carrier wave*. In order to carry information, it is modulated by speech or music. In amplitude modulation, the carrier amplitude is varied by the modulating signal, and this is shown in Fig. 1.5(a). In frequency modulation, the carrier frequency is varied by the modulating signal while the carrier amplitude remains constant, and this is shown in Fig. 1.5(b).

When the modulating signal consists of musical sounds, the frequency bandwidth of the music extends to about 10 kHz for commercial broadcasts or up to 15 kHz for high-fidelity music transmissions.

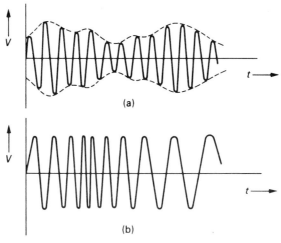

(a)

(b)

Fig. 1.5

Colour television signal

The colour television signal is designed to be compatible with both monochrome and colour receivers. A video carrier is used to carry the monochrome information and its amplitude varies according to the picture brightness, while a subcarrier is used to carry the colour intensity and hue according to the PAL system.[4]

The television signal shown in Fig. 1.6(a) consists of intervals of time during which the picture is transmitted with its varying shades of brightness. Interleaved with this are sets of pulses and a colour burst signal for synchronising the receiver circuits with that at the transmitter. A typical channel bandwidth used extends from d.c. level to 8 MHz for the BBC 625-line PAL system and is shown in Fig. 1.6(b).

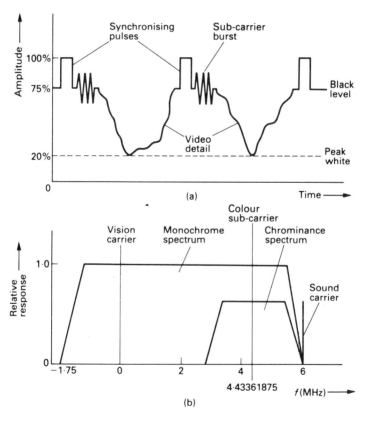

Fig. 1.6

Example 1.2
Discuss the frequency bandwidth required in line transmission when each of the following is transmitted
(a)   high-quality music,
(b)   12 telephone channels,
(c)   24 channels of voice-frequency telegraphy, each signalling at 50 bauds,
(d)   625-line television picture in which the picture line-scan occupies $60\,\mu s$ and the aspect ratio is $\frac{4}{3}$.   (U.L.)

Solution
(a)   High-quality music (Hi-Fi for short) would cover a frequency band over the range of average human audibility. This extends down to about 20 Hz and up to 16 or 18 kHz. The frequency bandwidth required is therefore approximately 16 kHz or the equivalent of four telephone channels as used in line transmission.
(b)   For intelligible speech a bandwidth of 3·4 kHz is sufficient, but to allow for filter characteristics an average of 4 kHz per channel is allocated. Hence, 12 channels require $12 \times 4\,\mathrm{kHz} = 48\,\mathrm{kHz}$ bandwidth.
(c)   Since the signalling speed in bauds equals the reciprocal of the duration of the telegraph pulse

$$\text{pulse duration} = \tfrac{1}{50}\ \text{s}$$

For a complete cycle, we require a double pulse change (+ ve to − ve) for the fundamental period. Hence

$$\text{duration of cycle} = 2 \times \tfrac{1}{50} = \tfrac{1}{25}\ \text{s}$$

or   fundamental frequency = 25 Hz

These pulses are used to amplitude-modulate a VF (voice-frequency) carrier tone of around 500 Hz. Sum and difference frequencies which are called sidebands are produced about the carrier tone to give an overall bandwidth of $2 \times 25\,\mathrm{Hz} = 50\,\mathrm{Hz}$.
However, to allow for filter characteristics, a bandwidth of 120 Hz is used in practice. It allows some of the third harmonic, i.e. 75 Hz, to be transmitted as well, which preserves the pulse squareness and so ensures the reliable operation of electromechanical relays (telegraph relays).
The bandwidth requirement for 24 channels is therefore $24 \times 120\,\mathrm{Hz} = 2880\,\mathrm{Hz}$.
(d)   Let the picture width be w and the height be h as shown in Fig. 1.7. Hence

$$w/h = \tfrac{4}{3}$$

or $h = 3w/4$

Since the picture has 625 lines, the distance between lines is given by

$$h/625 = 3w/(4 \times 625)$$

For equal horizontal and vertical resolution, the width of picture element is equal to its height, and so equals $3w/(4 \times 625)$. Hence

$$\text{number of elements per line} = w/\text{width of picture element} = \tfrac{4}{3} \times 625$$

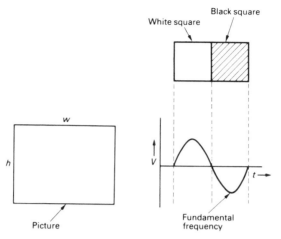

Fig. 1.7

Now

$$\text{time to scan a picture element} = \frac{60 \times 10^{-6}}{\frac{4}{3} \times 625} = \frac{9 \times 10^{-6}}{125} \text{ s}$$

or

$$\text{time to scan two picture elements (full cycle)} = \frac{18 \times 10^{-6}}{125} \text{ s}$$

Hence

$$\text{fundamental frequency} = \frac{125 \times 10^6}{18} = 7 \text{ MHz}$$

Since the average picture brightness is a d.c. signal, this must also be transmitted, and so the total bandwidth requirement is 0 to 7 MHz, i.e. 7 MHz. Because of this large bandwidth, double-sideband television is not used, but a more economical system called vestigial sideband transmission is used.

Pulsed radar signal

The location of distant targets in range and bearing by radar is usually achieved by transmitting a short periodic signal and receiving back some of the reflected signal from the target. This signal is basically a train of rectangular pulses, each pulse consisting of a short burst of RF energy, transmitted at a low repetition frequency of around 1 kHz and is shown in Fig. 1.8. The pulse width used varies between $0\cdot1\ \mu s$ and $10\ \mu s$ in duration.

Such a signal can be shown to consist of an infinite set of sine waves whose frequencies are harmonically related. The bandwidth required to receive such a signal with reasonable certainty, after reflection from a distant target, is about 2 to 5 MHz.

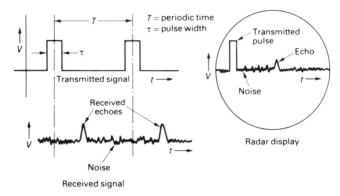

Fig. 1.8

Example 1.3

A radar transmitter is modulated at a repetition frequency of 1 kHz and transmits pulses of 4 μs duration. If the average power transmitted is 400 watts, determine

(a)   the duty ratio,
(b)   the peak power,
(c)   the minimum and maximum ranges.

Solution

(a) $$\text{duty ratio} = \frac{\text{pulse width}}{\text{periodic time}} = \frac{4 \times 10^{-6}}{1/1000} = \frac{1}{250}$$

(b) $$\text{average power} = \text{peak power} \times \text{duty ratio}$$

or $$\text{peak power} = \frac{400}{1/250} = 100 \text{ kW}$$

(c)   Electromagnetic energy travels at the velocity of 3×10^8 m/s.

Hence   distance travelled in 1 μs $= 3 \times 10^8 \times 10^{-6} = 300$ metres

The minimum distance is determined by the minimum time the signal takes to travel to the target and back again. This must not be less than the pulse duration or else the reflected signal will be confused with the transmitted signal which is only turned off after 4 μs.

Hence   minimum time to target $= \frac{4}{2}\mu\text{s} = 2 \mu\text{s}$

or   minimum distance to target $= 2 \times 300 = 600$ m

The maximum range is determined by the maximum time taken to the target and back again. This must be just equal to the time between the first and second pulses or else the reflected pulse will be confused with the second transmitted pulse.

Hence $\qquad$ maximum time to target $= \dfrac{\text{periodic time}}{2} = \dfrac{1}{2000} = 500\,\mu s$

or $\qquad$ maximum distance to target $= 500 \times 300 = 150\,\text{km}$

CW radar signal

A continuous-wave (CW) radar signal may be used for measuring the velocity of an aircraft. In Doppler radar, the transmitted and received signals are mixed together in a receiver to obtain the Doppler shift which is a measure of the velocity of the aircraft. Details of the Doppler effect are given in Appendix A.

Example 1.4

Explain how the Doppler effect is utilised in a radar system to measure aircraft velocity.

An aircraft approaches an airfield radar with a radial velocity of 800 km/h. If the radar frequency is 2 GHz, calculate the change in frequency of the received signal.

(C.E.I.)

Solution

The answer to the first part of the question will be found in Appendix A.

Problem

We have $\qquad$ $\Delta f = \dfrac{2vf \cos \theta}{c} = \dfrac{2vf}{c}$

with $\qquad$ $\Delta f = \dfrac{2 \times 800 \times 10^3 \times 2 \times 10^9}{60 \times 60 \times 3 \times 10^8}$

Hence $\qquad$ $\Delta f = \dfrac{8 \times 10^4}{27}$

or $\qquad$ $\Delta f \simeq 3\,\text{kHz}$

Sampled-data signal

The sampled-data signal shown in Fig. 1.9 is a particular example of a digital signal whose amplitude can vary over a continuous range of values. It is produced by 'sampling' an analogue signal at discrete instants of time. Impulse sampling or finite-width sampling may be used and, in the latter case, the signal is also called a pulse-amplitude modulated signal (PAM). Sampled-data signals are used extensively for digital signal processing in data systems.[6] Further details are given in Section 5.1.

1.3 Spectrum of a signal

Signals can be analysed by Fourier techniques into various frequency components. The total range of these frequencies represents the frequency spectrum of the signal and is of prime importance in telecommunications. An

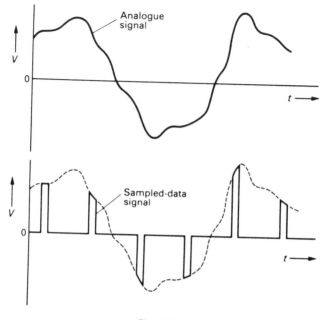

Fig. 1.9

exact knowledge of such a spectrum is useful in solving problems of transmission and reception.

A signal may thus be represented in the *time domain* as a plot of instantaneous amplitude against time or in the *frequency domain* as a plot of its spectral component amplitudes against frequency. There is a direct link between these two representations which is obtained with the aid of Fourier transforms.

It should be noted that there are two kinds of spectra called the discrete spectrum and the continuous spectrum; these are discussed in detail in the next chapter.

2

Signal analysis

2.1 Fourier series[7]

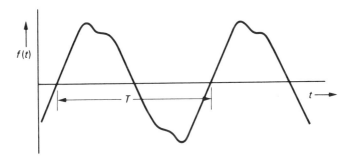

Fig. 2.1 A periodic signal

A signal which is repetitive is a periodic function of time. Any periodic function of time $f(t)$ can be represented by the Fourier series

$$f(t) = a_0 + \sum_1^\infty a_n \cos n\omega t + \sum_1^\infty b_n \sin n\omega t$$

where a_n and b_n are the coefficients to be evaluated and are given by the expressions

$$a_n = \frac{2}{T} \int_{-T/2}^{+T/2} f(t) \cos n\omega t \, dt$$

$$b_n = \frac{2}{T} \int_{-T/2}^{+T/2} f(t) \sin n\omega t \, dt$$

where $\omega = 2\pi/T$ and T is the periodic time.

The d.c. term is a_0 and is given by the average value of $f(t)$ in a period T.

$$a_0 = \frac{1}{T} \int_{-T/2}^{+T/2} f(t) \, dt$$

Comments
1. If $f(t) = f(-t)$, the function is even. There is symmetry about the origin and only cosine terms are present (d.c. term optional).
2. If $f(t) = -f(-t)$, the function is odd and only sine terms are present (d.c. term optional).
3. If $f(t + T/2) = f(t)$, only even harmonics are present.
4. If $f(t + T/2) = -f(t)$, only odd harmonics are present.

2.2 Discrete spectrum

The Fourier series represents an infinite number of frequency components which added together yield the time function $f(t)$. These frequency components constitute a *discrete* spectrum which is shown in Fig. 2.2 and the amplitudes of each discrete frequency are given by the coefficients a_n and b_n. All the frequency components are harmonics of the fundamental frequency $1/T$ and the total range of the frequencies is the bandwidth of the signal.

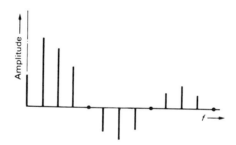

Fig. 2.2 Discrete spectrum

Though the frequency spectrum may consist of an infinite number of discrete frequencies, their amplitudes get smaller with larger values of n and, in practice, it is sufficient to consider only a finite number of the frequencies as adequate for communications.

The significance of this lies in the need to economise in the use of bandwidth in communication systems. A knowledge of the frequency spectrum will assist the transmission and reception of the signal in the most effective and economical manner.

Example 2.1
The square-wave signal shown in Fig. 2.3 varies between the values of $+1$ and -1. It has a periodic time T and is symmetrical with respect to the vertical axis at time $t = 0$. Obtain the Fourier components of the waveform.

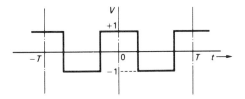

Fig. 2.3

Solution

The waveform is also symmetrical about the horizontal axis, so the average area is zero. Hence, the d.c. term is zero, i.e. $a_0 = 0$. In addition, $f(t) = f(-t)$ and so only cosine terms are present, i.e. $b_n = 0$.

Now
$$a_n = \frac{2}{T} \int_{-T/2}^{+T/2} f(t) \cos n\omega t \, dt$$

where
$$f(t) = -1 \quad \text{from } -T/2 \text{ to } -T/4$$
$$f(t) = +1 \quad \text{from } -T/4 \text{ to } +T/4$$
$$f(t) = -1 \quad \text{from } +T/4 \text{ to } +T/2$$

Hence

$$a_n = \frac{2}{T} \left\{ \int_{-T/2}^{-T/4} -\cos n\omega t \, dt + \int_{-T/4}^{T/4} \cos n\omega t \, dt - \int_{T/4}^{T/2} \cos n\omega t \, dt \right\}$$

$$= \frac{2}{T} \left\{ \left[\frac{-\sin n\omega t}{n\omega} \right]_{-T/2}^{-T/4} + \left[\frac{\sin n\omega t}{n\omega} \right]_{-T/4}^{T/4} - \left[\frac{\sin n\omega t}{n\omega} \right]_{T/4}^{T/2} \right\}$$

$$= \frac{2}{n\omega T} \left\{ -\sin\left(\frac{-n\omega T}{4}\right) + \sin\left(\frac{-n\omega T}{2}\right) + \sin\left(\frac{n\omega T}{4}\right) - \sin\left(\frac{-n\omega T}{4}\right) \right.$$
$$\left. - \sin\left(\frac{n\omega T}{2}\right) + \sin\left(\frac{n\omega T}{4}\right) \right\}$$

$$= \frac{8}{n\omega T} \sin\left(\frac{n\omega T}{4}\right) - \frac{4}{n\omega T} \sin\left(\frac{n\omega T}{2}\right)$$

With $\omega T = 2\pi$, the second term is zero for all integer values of n. Hence

$$a_n = \frac{8}{2n\pi} \sin\left(\frac{n\pi}{2}\right) = \frac{4}{n\pi} \sin\left(\frac{n\pi}{2}\right)$$

$$a_0 = 0 \quad \text{(d.c. term)}$$

$$a_1 = \frac{4}{\pi} \sin\left(\frac{\pi}{2}\right) = \frac{4}{\pi}$$

$$a_2 = \frac{4}{2\pi}\sin \pi = 0$$

$$a_3 = \frac{4}{3\pi}\sin\left(\frac{3\pi}{2}\right) = \frac{-4}{3\pi}$$

. . . .

Now $\qquad f(t) = a_0 + a_1\cos\omega t + a_2\cos 2\omega t + \dots$

or $\qquad f(t) = \frac{4}{\pi}(\cos\omega t - \tfrac{1}{3}\cos 3\omega t + \tfrac{1}{5}\cos 5\omega t - \dots)$

2.3 Typical series

The Fourier components of typical waveforms used in practice are shown in Figs. 2.4 to 2.7.

Symmetrical square wave

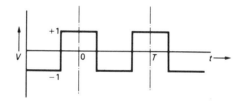

Fig. 2.4

$$f(t) = \frac{4}{\pi}(\cos\omega t - \tfrac{1}{3}\cos 3\omega t + \tfrac{1}{5}\cos 5\omega t - \dots)$$

Asymmetrical square wave

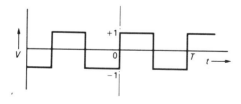

Fig. 2.5

$$f(t) = \frac{4}{\pi}(\sin\omega t + \tfrac{1}{3}\sin 3\omega t + \tfrac{1}{5}\sin 5\omega t + \dots)$$

Triangular wave

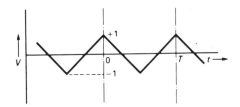

Fig. 2.6

$$f(t) = \frac{8}{\pi^2}(\cos \omega t + \tfrac{1}{9}\cos 3\omega t + \tfrac{1}{25}\cos 5\omega t + \ldots)$$

Sawtooth wave

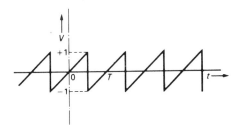

Fig. 2.7

$$f(t) = \frac{2}{\pi}(\sin \omega t - \tfrac{1}{2}\sin 2\omega t + \tfrac{1}{3}\sin 3\omega t - \ldots)$$

2.4 Complex form

An alternative but convenient way of writing the periodic function $f(t)$ is in terms of complex quantities. From complex signal theory, we have the relationships

$$\cos n\omega t = \frac{e^{jn\omega t} + e^{-jn\omega t}}{2}$$

$$\sin n\omega t = \frac{e^{jn\omega t} - e^{-jn\omega t}}{2j}$$

Substituting in the expression for the Fourier series gives

$$f(t) = a_0 + \sum_1^\infty a_n \left(\frac{e^{jn\omega t} + e^{-jn\omega t}}{2} \right) + \sum_1^\infty b_n \left(\frac{e^{jn\omega t} - e^{-jn\omega t}}{2j} \right)$$

$$= a_0 + \sum_1^\infty \left\{ \frac{(a_n - jb_n)e^{jn\omega t}}{2} + \frac{(a_n + jb_n)e^{-jn\omega t}}{2} \right\}$$

Put $C_n = \frac{1}{2}(a_n - jb_n)$
$C_{-n} = \frac{1}{2}(a_n + jb_n)$
$C_0 = a_0$

where C_{-n} is the complex conjugate of C_n. Substituting expressions for the coefficients a_n and b_n from Section 2.1 yields

$$C_n = \frac{1}{T} \int_{-T/2}^{+T/2} f(t) [\cos n\omega t - j \sin n\omega t] \, dt$$

$$C_{-n} = \frac{1}{T} \int_{-T/2}^{+T/2} f(t) [\cos n\omega t + j \sin n\omega t] \, dt$$

or

$$C_n = \frac{1}{T} \int_{-T/2}^{+T/2} f(t) e^{-jn\omega t} \, dt$$

and

$$C_{-n} = \frac{1}{T} \int_{-T/2}^{+T/2} f(t) e^{jn\omega t} \, dt$$

with

$$f(t) = C_0 + \sum_1^\infty C_n e^{jn\omega t} + \sum_{-\infty}^{-1} C_n e^{jn\omega t}$$

where the values of n are negative in the last term and are included with the sigma sign. Furthermore, C_0 can be included under the sigma sign by using the value $n = 0$ also. Hence

$$f(t) = \sum_{-\infty}^{+\infty} C_n e^{jn\omega t}$$

This result shows that the periodic function $f(t)$ can also be represented mathematically by an infinite set of positive and negative frequency components. The negative frequencies have a mathematical significance and they may sometimes also have physical significance, as a positive frequency can be associated with an anticlockwise rotation and a negative frequency with a clockwise rotation.

Example 2.2

Deduce the Fourier series which corresponds to the waveform of a positive-going rectangular pulse train, each pulse of duration τ, repetitive in the period T, and with an amplitude E. Show that harmonics at frequencies which are multiples of $1/\tau$ have zero

amplitude. Sketch the waveform and state a corresponding expression for the output from each of the following ideal filters when the pulse train is applied to the input
(a) high-pass, cut-off frequency $1/2T$,
(b) low-pass, cut-off frequency $1/2T$,
(c) low-pass, cut-off frequency $3/2T$,
(d) band-pass, cut-off frequencies $1/2T$ and $3/2T$,
(e) band-stop, cut-off frequencies $1/2T$ and $3/2T$. (U.L.)

Fig. 2.8 Pulse train

Solution
The general Fourier series for the pulse train above is given by

$$f(t) = a_0 + \sum_1^\infty a_n \cos n\omega t + \sum_1^\infty b_n \sin n\omega t$$

where

$$a_0 = \frac{1}{T}\int_{-T/2}^{+T/2} f(t)\,dt = \frac{1}{T}\int_{-\tau/2}^{+\tau/2} E\,dt = \frac{E}{T}[t]_{-\tau/2}^{+\tau/2} = \frac{E\tau}{T}$$

Here, the b_n coefficients are zero due to the *choice* of $t = 0$ at the centre of a pulse. The a_n coefficients are given by

$$a_n = \frac{2}{T}\int_{-T/2}^{+T/2} f(t)\cos n\omega t\,dt = \frac{2}{T}\int_{-\tau/2}^{+\tau/2} E\cos n\omega t\,dt$$

$$= \frac{2E}{T}\left[\frac{\sin n\omega t}{n\omega}\right]_{-\tau/2}^{+\tau/2} = \frac{2E}{n\omega T}\left[\sin\frac{n\omega\tau}{2} - \sin\frac{-n\omega\tau}{2}\right]$$

or

$$a_n = \frac{4E}{n\omega T}\sin\frac{n\omega\tau}{2}$$

and

$$f(t) = \frac{E\tau}{T} + \frac{2E\tau}{T}\sum_1^\infty \frac{\sin(n\omega\tau/2)}{n\omega\tau/2}\cos n\omega t$$

This is shown in Fig. 2.9.
   The envelope of components other than a_0 is of the form $(\sin x)/x$ where $x = n\omega\tau/2$ and the zeros occur when

$$\frac{\sin x}{x} = 0$$

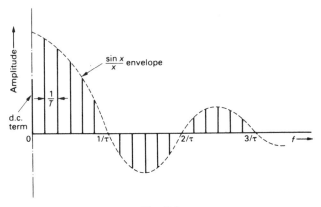

Fig. 2.9

or $\qquad n\omega\tau/2 = k\pi \quad$ where $k = 1, 2, 3,$ etc.

and $\qquad nf = k/\tau \quad$ which are multiples of $1/\tau$

Hence, the zeros occur at $1/\tau$, $2/\tau$, etc.

(a)   All frequencies except the d.c. term:

$$f(t) = \frac{2E\tau}{T} \sum_{1}^{\infty} \frac{\sin x}{x} \cos n\omega t$$

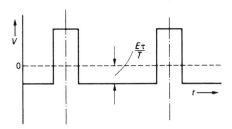

Fig. 2.10

(b)   The d.c. term only:

$$f(t) = \frac{E\tau}{T}$$

Fig. 2.11

(c) The d.c. term and first harmonic only:

$$f(t) = \frac{E\tau}{T} + \frac{2E\tau}{T}\frac{\sin x}{x}\cos \omega t$$

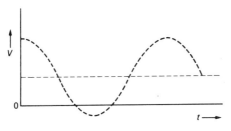

Fig. 2.12

(d) The first harmonic only:

$$f(t) = \frac{2E\tau}{T}\frac{\sin x}{x}\cos \omega t$$

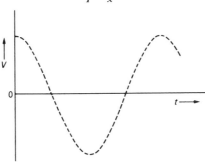

Fig. 2.13

(e)   All the components except the fundamental:

$$f(t) = \frac{E\tau}{T} + \sum_2^\infty \frac{2E\tau}{T} \frac{\sin x}{x} \cos n\omega t$$

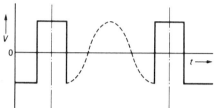

Fig. 2.14

Example 2.3

The periodic pulse train shown in Fig. 2.15 can be expressed by the Fourier series

$$f(t) = E\left[\frac{1}{6} + \frac{1}{\pi}\cos(2\pi \times 10^4)t + \sum_2^\infty \frac{2}{\pi n}\sin\frac{(n\pi)}{6}\cos(2\pi n \times 10^4)t\right]$$

Deduce the corresponding series when the pulse train is amplitude-modulated to a depth of 50% by a sinusoidal signal with a frequency of 3 kHz. Indicate the main components of the frequency spectrum produced.

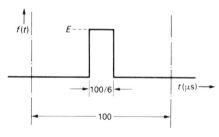

Fig. 2.15

Calculate the rms value and sketch the waveform of the voltage at the output of
(a)   a low-pass filter with a cut-off frequency at 4 kHz,
(b)   a band-pass filter with cut-off frequencies at 30 kHz ± 4 kHz when the modulated pulse train is applied to the input and $E = 12$ V.   (U.L.)

Solution
The expression for the unmodulated train when $E = 12$ V is

$$f(t) = 2 + \frac{12}{\pi}\cos(2\pi \times 10^4)t + \frac{6\sqrt{3}}{\pi}\cos(4\pi \times 10^4)t + \frac{8}{\pi}\cos(6\pi \times 10^4)t + \ldots$$

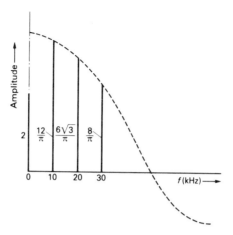

Fig. 2.16

The frequency spectrum is plotted in Fig. 2.16 up to the third harmonic.

Let the modulating signal be $A \sin \omega_a t$ where $\omega_a = 2\pi \times 3 \times 10^3$. The expression for the modulated train $F(t)$ is given by

$$F(t) = (1 + m \sin \omega_a t)\left[2 + \frac{12}{\pi} \cos (2\pi \times 10^4)t + \frac{6\sqrt{3}}{\pi} \cos (4\pi \times 10^4)t \right.$$

$$\left. + \frac{8}{\pi} \cos (6\pi \times 10^4)t + \ldots \right]$$

where $m = A/E = 0.5$ is the depth of modulation.

The spectrum of the modulated pulse train will consist of the original frequency components, and about each component there will be a pair of sidebands spaced $\pm 3\,\mathrm{kHz}$ on either side. The amplitude of the sidebands is $m/2$ times that of the corresponding harmonic, while the amplitude of the 3 kHz component is given by $m \times 2\,\mathrm{V} = \frac{1}{2} \times 2\,\mathrm{V} = 1\,\mathrm{V}$. All amplitudes are peak values and the spectrum is plotted in Fig. 2.17.

(a)   The d.c. component and the 3 kHz component:

$$\text{rms value} = \sqrt{2^2 + (1/\sqrt{2})^2} = \sqrt{4.5} = 2.12\,\mathrm{V}$$

The waveform is shown in Fig. 2.18.

(b)   Three frequencies of 27 kHz, 30 kHz, and 33 kHz are filtered out:

$$\text{rms value} = \sqrt{(2/\pi\sqrt{2})^2 + (8/\pi\sqrt{2})^2 + (2/\pi\sqrt{2})^2} = \frac{\sqrt{36}}{\pi} = 1.91\,\mathrm{V}$$

The waveform is shown in Fig. 2.19.

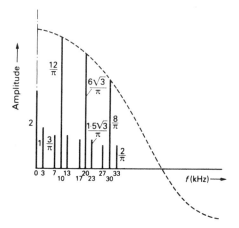

Fig. 2.17

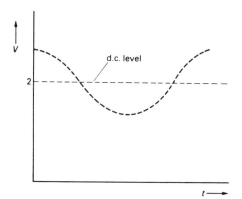

Fig. 2.18

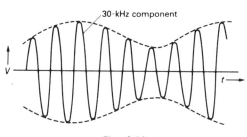

Fig. 2.19

2.5 Fourier transform

The Fourier series technique can be extended to non-periodic waveforms such as single pulses or a single transient by making $T \to \infty$. In this way, adjacent pulses virtually never occur and the pulse train reduces to a single isolated pulse. Assuming $f(t)$ is *initially* periodic, we have

$$f(t) = \sum_{-\infty}^{+\infty} C_n e^{jn\omega t}$$

where

$$C_n = \frac{1}{T} \int_{-T/2}^{+T/2} f(t) e^{-jn\omega t} \, dt$$

In the limit, for a single pulse we have

$$T \to \infty \qquad \omega = \frac{2\pi}{T} \to d\omega \quad \text{(a small quantity)}$$

or

$$\frac{1}{T} = \frac{\omega}{2\pi} \to \frac{d\omega}{2\pi}$$

Furthermore, the nth harmonic in the Fourier series which is $n\omega \to n\, d\omega$ and this becomes some general value which will be defined as 'ω'.

In the limit, the sigma sign leads to an integral and we have

$$C_n = \frac{d\omega}{2\pi} \int_{-\infty}^{+\infty} f(t) e^{-j\omega t} \, dt$$

and

$$f(t) = \int_{-\infty}^{+\infty} \frac{d\omega}{2\pi} \left[\int_{-\infty}^{+\infty} f(t) e^{-j\omega t} \, dt \right] e^{j\omega t}$$

The quantity in brackets, when evaluated, is a function of frequency only and is denoted as $F(\omega)$ where

$$F(\omega) = \int_{-\infty}^{+\infty} f(t) e^{-j\omega t} \, dt$$

It is called the Fourier transform of $f(t)$. Substituting for $f(t)$ above, we obtain

$$f(t) = \frac{1}{2\pi} \int_{-\infty}^{+\infty} F(\omega) e^{j\omega t} \, d\omega$$

which is called the inverse Fourier transform. The time function $f(t)$ now represents the expression for a *single* pulse or transient only.

2.6 Continuous spectrum

The significance of this final result is that any single pulse or transient can be expressed as the sum of an infinite number of frequency components $F(\omega)$ where ω is any general value.

This leads to a *continuous* spectrum in contrast to the discrete spectrum of the periodic waveform. In physical terms, the frequency components are all crowded very close to one another because the spacing between them is $1/T$ which tends to zero as $T \to \infty$.

In general, $F(\omega)$ is complex and its amplitude and phase can be plotted to give the frequency spectrum of the time function $f(t)$. An example of this is shown in the next section for a single rectangular pulse centred at $t = 0$.

The quantity $|F(\omega)|$, when plotted, shows a variation of amplitude against ω and so the quantity $|F(\omega)| \, d\omega$ represents an elementary area of this graph, within a range $d\omega$, and is called the *spectral density*.

2.7 Typical functions

(a) Rectangular pulse A

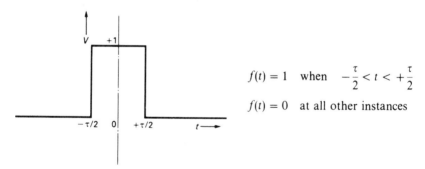

$$f(t) = 1 \quad \text{when} \quad -\frac{\tau}{2} < t < +\frac{\tau}{2}$$

$$f(t) = 0 \quad \text{at all other instances}$$

Fig. 2.20

$$F(\omega) = \int_{-\infty}^{-\tau/2} f(t)\,e^{-j\omega t}\,dt + \int_{-\tau/2}^{+\tau/2} f(t)\,e^{-j\omega t}\,dt + \int_{+\tau/2}^{+\infty} f(t)\,e^{-j\omega t}\,dt$$

or
$$F(\omega) = 0 + \int_{-\tau/2}^{+\tau/2} e^{-j\omega t}\,dt + 0$$

$$= \left[\frac{e^{-j\omega t}}{-j\omega}\right]_{-\tau/2}^{+\tau/2} = \frac{1}{j\omega}(e^{j\omega\tau/2} - e^{-j\omega\tau/2})$$

Hence $$F(\omega) = \frac{2}{\omega} \sin\left(\frac{\omega\tau}{2}\right) = \tau \frac{\sin(\omega\tau/2)}{\omega\tau/2}$$

or $$F(\omega) = \tau \frac{\sin x}{x} \quad \text{where} \quad x = \frac{\omega\tau}{2}$$

with $$\frac{F(\omega)}{\tau} = \frac{\sin x}{x}$$

A plot of $F(\omega)/\tau$ is the familiar $(\sin x)/x$ curve and is shown in Fig. 2.21. It is a *continuous* curve and is symmetrical in x, i.e. its value is unchanged when x becomes negative.

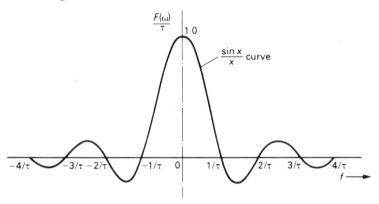

Fig. 2.21 Continuous spectrum

Alternatively, the amplitude and phase can be drawn separately as in Fig. 2.22.

Comments

1. $\lim\limits_{x \to 0} \dfrac{\sin x}{x} = 1$. The peak value of the graph is one.

2. The zeros occur when $\dfrac{\sin x}{x} = 0$.

   Hence $$\sin x = 0 \quad \text{or} \quad x = n\pi \quad (n = 1, 2, \text{ etc.} \ldots)$$

   and $$\frac{\omega\tau}{2} = \pi, 2\pi, \ldots$$

   or $$f = \frac{1}{\tau}, \frac{2}{\tau}, \ldots$$

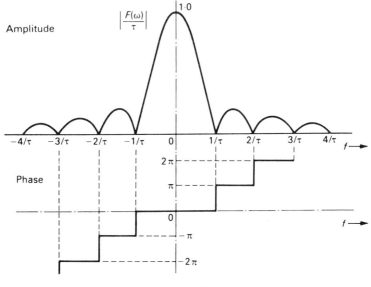

Fig. 2.22

3. The phase shifts by π radians when the graph changes polarity from positive to negative or vice versa.

(b) Rectangular pulse B

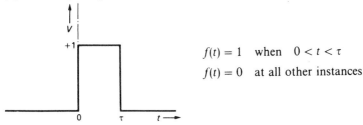

$f(t) = 1$ when $0 < t < \tau$

$f(t) = 0$ at all other instances

Fig. 2.23

$$F(\omega) = 0 + \int_0^{\tau} e^{-j\omega t} \, dt + 0$$

$$= \left[\frac{e^{-j\omega t}}{-j\omega} \right]_0^{\tau} = \frac{e^{-j\omega\tau} - 1}{-j\omega}$$

$$= \frac{e^{-j\omega(\tau/2)}}{-j\omega} (e^{-j\omega(\tau/2)} - e^{j\omega(\tau/2)})$$

Hence
$$F(\omega) = \frac{2e^{-j\omega(\tau/2)}}{\omega} \left(\frac{e^{j\omega(\tau/2)} - e^{-j\omega(\tau/2)}}{2j} \right)$$

$$= \tau\, e^{-j\omega(\tau/2)} \frac{\sin(\omega\tau/2)}{\omega\tau/2}$$

$$= \tau\, \frac{\sin x}{x} e^{-jx} \quad \text{where} \quad x = \frac{\omega\tau}{2}$$

or
$$\frac{F(\omega)}{\tau} = \frac{\sin x}{x} e^{-jx}$$

The amplitude and phase are plotted in Fig. 2.24.

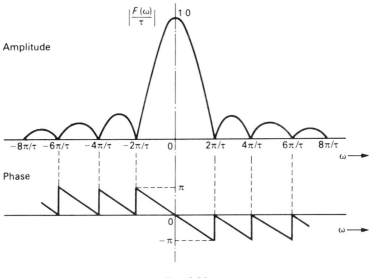

Fig. 2.24

Comments
1. The shape of the spectrum is exactly the same as that of the previous case.
2. There is an additional uniform phase shift factor e^{-jx} which alters the phase diagram of the previous case.
3. The result above can readily be obtained by using the shift theorem given below.

Shift theorem
The Fourier transform of any time function $f(t)$ delayed by τ_0 is simply the same Fourier transform delayed by a phase factor $e^{-j\omega\tau_0}$.

Proof
Let $F_1(\omega)$ be the Fourier transform of $f(t)$ and $F_2(\omega)$ be that of $f(t - \tau_0)$.

Hence
$$F_2(\omega) = \int_{-\infty}^{+\infty} f(t - \tau_0) e^{-j\omega t} \, dt$$

or
$$F_2(\omega) = \int_{-\infty}^{+\infty} f(t') e^{-j\omega t'} e^{-j\omega \tau_0} \, dt'$$

where $t' = t - \tau_0$, i.e. $t = t' + \tau_0$, and $dt = dt'$ ($\tau_0 =$ a constant).

But
$$\int_{-\infty}^{+\infty} f(t') e^{-j\omega t'} \, dt' \equiv \int_{-\infty}^{+\infty} f(t) e^{-j\omega t} \, dt = F_1(\omega)$$

because t' is any value of t.

Hence
$$F_2(\omega) = F_1(\omega) e^{-j\omega \tau_0}$$

(c)   Impulse function (unit impulse)

It is also called the Dirac delta function $\delta(t)$ which theoretically has an infinite amplitude and is infinitely narrow. It is defined by

1. $\delta(t) = 0$ for all values except at $t = 0$.
2. $\int_{-\infty}^{+\infty} \delta(t) \, dt = 1$.

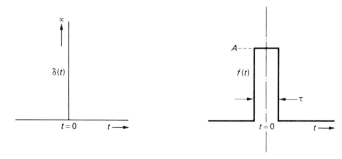

Fig. 2.25

The second condition implies that the pulse exists in the neighbourhood of $t = 0$ and the area within it is unity. A similar result is obtained by considering the delta function as the limit of a finite rectangular pulse $f(t)$ of amplitude A and duration τ, if the area $A\tau$ within it is unity.

Hence
$$\delta(t) = \lim_{\substack{\tau \to 0 \\ A \to \infty}} f(t) \quad \text{where} \quad f(t) = A \quad \text{for} \quad -\frac{\tau}{2} < t < \frac{\tau}{2}$$
$$f(t) = 0 \quad \text{for all other values}$$

If the Fourier transform of $\delta(t)$ is $F(\omega)$, then

$$F(\omega) = \int_{-\infty}^{+\infty} \left[\lim_{\substack{\tau \to 0 \\ A \to \infty}} f(t) \right] e^{-j\omega t}\, dt$$

$$= \lim_{\substack{\tau \to 0 \\ A \to \infty}} \int_{-\infty}^{+\infty} f(t)\, e^{-j\omega t}\, dt$$

The quantity under the integral sign is the Fourier transform of a rectangular pulse of amplitude A as in Section 2.7(a) above.

Hence $\qquad F(\omega) = \lim_{\substack{\tau \to 0 \\ A \to \infty}} \left[A\tau \frac{\sin(\omega\tau/2)}{\omega\tau/2} \right] = \lim_{\substack{\tau \to 0 \\ A \to \infty}} A\tau \lim_{\tau \to 0} \frac{\sin(\omega\tau/2)}{\omega\tau/2}$

or $\qquad\qquad\qquad\qquad F(\omega) = 1 \times 1 = 1$

The frequency spectrum which is shown in Fig. 2.26 has a constant amplitude and extends over all values of ω, positive or negative.

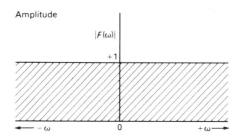

Fig. 2.26

(d) Step function
This is the well-known driving function and is obtained by suddenly closing a switch in a d.c. circuit. The analysis is made easier by considering the function to be made up of two component waveforms as illustrated in Fig. 2.27.

The first waveform is similar to the Signum function shown in Fig. 2.28(a) but has half its amplitude. From Appendix B, its Fourier transform is given by

$$F_1(\omega) = \frac{1}{2}\left(\frac{2}{j\omega} \right) = \frac{1}{j\omega}$$

The second waveform can be associated with the unit impulse function shown in Fig. 2.28(b). From Appendix B, its Fourier transform is given by

$$F_2(\omega) = \pi\delta(\omega)$$

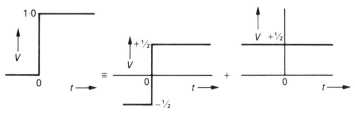

Fig. 2.27

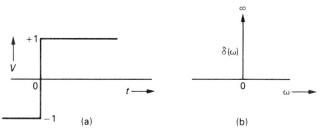

Fig. 2.28

Hence, the Fourier transform of the step function becomes

$$F(\omega) = F_1(\omega) + F_2(\omega)$$

or

$$F(\omega) = (1/j\omega) + \pi\delta(\omega)$$

Example 2.4

What essential property of a time function determines whether it will possess a line spectrum or a continuous spectrum? Explain fully what is meant by both of these terms.
  Determine the frequency spectrum of a single pulse specified by the conditions

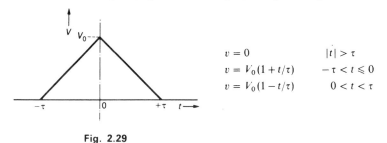

$$
\begin{aligned}
v &= 0 & |t| &> \tau \\
v &= V_0(1 + t/\tau) & -\tau &< t \leqslant 0 \\
v &= V_0(1 - t/\tau) & 0 &< t < \tau
\end{aligned}
$$

Fig. 2.29

where t represents time and τ is a constant. Sketch the spectrum obtained and briefly discuss its properties.

Write down, but do not attempt to evaluate, the expression for the pulse shape which would result if the above signal were passed through a filter which cut off all frequencies greater than $2\pi/\tau$, but left all other frequencies unaffected. (U.L.)

Solution
A *periodic* time function yields a line spectrum, while a *non-periodic* time function yields a continuous spectrum. In the former case only discrete frequencies, which are harmonically related, exist, while in the latter case all the frequencies exist to give a continuous distribution from 0 to ∞.

Problem

$$F(\omega) = \int_{-\infty}^{+\infty} f(t)\,e^{-j\omega t}\,dt$$

$$= \int_{-\tau}^{0} V_0(1 + t/\tau)\,e^{-j\omega t}\,dt + \int_{0}^{\tau} V_0(1 - t/\tau)\,e^{-j\omega t}\,dt$$

or

$$F(\omega) = V_0 \left\{ \int_{-\tau}^{0} e^{-j\omega t}\,dt + \frac{1}{\tau}\int_{-\tau}^{0} t\,e^{-j\omega t}\,dt \right.$$

$$\left. + \int_{0}^{\tau} e^{-j\omega t}\,dt - \frac{1}{\tau}\int_{0}^{\tau} t\,e^{-j\omega t}\,dt \right\}$$

$$= V_0 \left\{ \left[\frac{e^{-j\omega t}}{-j\omega} \right]_{-\tau}^{0} + \frac{1}{\tau}\left[e^{-j\omega t}\left(\frac{1}{\omega^2} - \frac{t}{j\omega} \right) \right]_{-\tau}^{0} \right.$$

$$\left. + \left[\frac{e^{-j\omega t}}{-j\omega} \right]_{0}^{\tau} - \frac{1}{\tau}\left[e^{-j\omega t}\left(\frac{1}{\omega^2} - \frac{t}{j\omega} \right) \right]_{0}^{\tau} \right\}$$

$$= V_0 \left\{ -\frac{1}{j\omega} + \frac{e^{j\omega\tau}}{j\omega} + \frac{1}{\omega^2\tau} - \frac{e^{j\omega\tau}}{\omega^2\tau} - \frac{e^{j\omega\tau}}{j\omega} \right.$$

$$\left. - \frac{e^{-j\omega\tau}}{j\omega} + \frac{1}{j\omega} - \frac{e^{-j\omega\tau}}{\omega^2\tau} + \frac{e^{-j\omega\tau}}{j\omega} + \frac{1}{\omega^2\tau} \right\}$$

$$= V_0 \left\{ \frac{2}{\omega^2\tau} - \frac{2}{\omega^2\tau}\frac{(e^{j\omega\tau} + e^{-j\omega\tau})}{2} \right\}$$

$$= \frac{2V_0}{\tau} \left\{ \frac{1}{\omega^2}(1 - \cos\omega\tau) \right\} = \frac{4V_0\tau}{\omega^2\tau^2}\sin^2\left(\frac{\omega\tau}{2} \right)$$

Hence

$$F(\omega) = V_0\tau \left(\frac{\sin x}{x} \right)^2 \quad \text{where} \quad x = \frac{\omega\tau}{2}$$

The spectrum is a $(\sin x)/x$ curve *squared*, which gives a more peaky waveform than that due to $(\sin x)/x$ only. Moreover, all areas are positive and most of the energy lies in the centre peak between $-1/\tau$ and $+1/\tau$. The curve is shown in Fig. 2.30.

Impulse theorem
Certain time functions can be reduced to a set of impulses by repeated differentiation. If $F(\omega)$ is the Fourier transform of $f(t)$, then differentiating $f(t)$

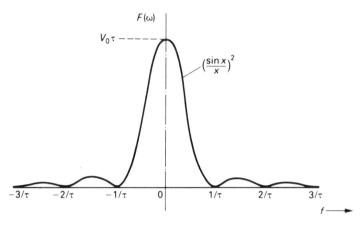

Fig. 2.30

n times gives

$$\frac{d^n}{dt^n}[f(t)] = \frac{d^n}{dt^n}\frac{1}{2\pi}\int_{-\infty}^{+\infty} F(\omega)\,e^{j\omega t}\,d\omega$$

$$= (j\omega)^n \frac{1}{2\pi}\int_{-\infty}^{+\infty} F(\omega)\,e^{j\omega t}\,d\omega$$

or $\qquad \mathscr{F}\left[\dfrac{d^n}{dt^n}[f(t)]\right] = (j\omega)^n F(\omega)$

When the quantity in brackets is a set of impulses whose Fourier transform is known (see Section 2.7(c)), then $F(\omega)$ can be determined.

As an illustration of this technique, consider $f(t)$ in the previous example. The waveforms for $f(t)$, $f'(t)$, and $f''(t)$ are shown in Fig. 2.31.

Here $\qquad \dfrac{d^2}{dt^2}[f(t)] = f''(t) = \dfrac{1}{\tau}\delta(t+\tau) + \dfrac{1}{\tau}\delta(t-\tau) - \dfrac{2}{\tau}\delta(t)$

and $\qquad \mathscr{F}[f''(t)] = \dfrac{1}{\tau}(e^{j\omega\tau} + e^{-j\omega\tau} - 2)$

Hence $\qquad F(\omega) = \dfrac{1}{(j\omega)^2}\dfrac{1}{\tau}(e^{j\omega\tau} + e^{-j\omega\tau} - 2)$

or $\qquad F(\omega) = \dfrac{2}{\omega^2\tau}(1 - \cos\omega\tau) = \dfrac{\tau\sin^2(\omega\tau/2)}{\omega^2\tau^2/4}$

which is the same result as obtained previously, if $V_0 = 1$ V.

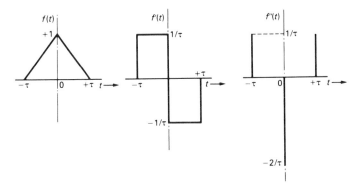

Fig. 2.31

Example 2.5

Use the Fourier integral

$$F(\omega) = \int_{-\infty}^{+\infty} f(t) e^{-j\omega t} dt$$

to derive expressions for the spectral amplitudes $|F(\omega)|$ corresponding to the waveforms of the pulse (a) and the double pulse (b) shown in Fig. 2.32.

Sketch the spectral distributions and comment on the difference of the low-frequency content with respect to transmission systems. (U.L.)

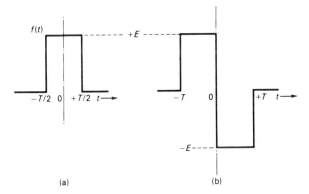

Fig. 2.32

Solution

(a) The solution is given in Section 2.7(a) by replacing τ with T and increasing the pulse amplitude to E.

Hence
$$F(\omega) = E T \frac{\sin(\omega T/2)}{\omega T/2} = E T \frac{\sin x}{x} \quad \text{where} \quad x = \frac{\omega T}{2}$$

The zeros occur when $(\sin x)/x = 0$, i.e. $\sin x = 0$

or
$$\omega T/2 = k\pi \quad (k = 1, 2, 3, \text{etc.} \dots)$$

and $f = k/T$ which are multiples of $1/T$.
The spectrum is shown in Fig. 2.33.

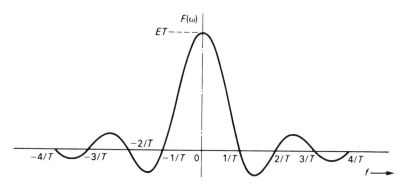

Fig. 2.33

(b)
$$F(\omega) = \int_{-\infty}^{+\infty} f(t) e^{-j\omega t} \, dt$$

$$= \int_{-T}^{0} E e^{-j\omega t} \, dt + \int_{0}^{+T} - E e^{-j\omega t} \, dt$$

$$= \frac{E}{-j\omega} \left[e^{-j\omega t} \right]_{-T}^{0} - \frac{E}{-j\omega} \left[e^{-j\omega t} \right]_{0}^{+T}$$

$$= -\frac{E}{j\omega} (1 - e^{j\omega T}) + \frac{E}{j\omega} (e^{-j\omega T} - 1)$$

$$= \frac{E}{j\omega} (e^{j\omega T} + e^{-j\omega T} - 2)$$

$$= \frac{2E}{j\omega} (\cos \omega T - 1) = j \frac{2E}{\omega} (1 - \cos \omega T)$$

Hence
$$F(\omega) = j\omega E T^2 \left(\frac{\sin x}{x} \right)^2 \quad \text{where} \quad x = \frac{\omega T}{2}$$

or
$$|F(\omega)| = \omega E T^2 \left(\frac{\sin x}{x} \right)^2 = E T 2x \left(\frac{\sin x}{x} \right)^2$$

The spectrum is plotted in Fig. 2.34.

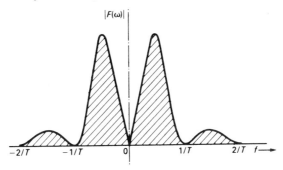

Fig. 2.34

Comment

In pulse (a), most of the energy is contained in the lower frequencies around the d.c. level and falls off at the higher frequencies near $1/T$.

In pulse (b), there is no d.c. component and energy is also mainly in the lower frequencies with a falling off at the higher frequencies near $1/T$.

Hence, in both cases, most of the energy is in the lower frequencies and within a bandwidth of $1/T$. The pulses are used for signalling in transmission systems; pulse (b) is especially used because no d.c. power is consumed. This pulse can also be used to time-multiplex two channels using the same overall bandwidth $1/T$, which is known as duobinary signalling.[8]

2.8 Transform pairs

A summary of the waveforms of typical transform pairs is given in Fig. 2.35.

2.9 Power and energy spectra

Signals are usually treated in terms of a voltage amplitude, but it is sometimes useful to know the power or energy associated with a signal such as noise.

Power spectrum

In the case of a periodic signal, the average power associated with a voltage $f(t)$ in a resistance of 1 ohm is given by

$$P_{av} = \frac{1}{T} \int_{-T/2}^{+T/2} f^2(t) \, dt = \frac{1}{T} \int_{-T/2}^{+T/2} f(t) \sum_{-\infty}^{+\infty} C_n e^{jn\omega t} \, dt$$

from Section 2.4.

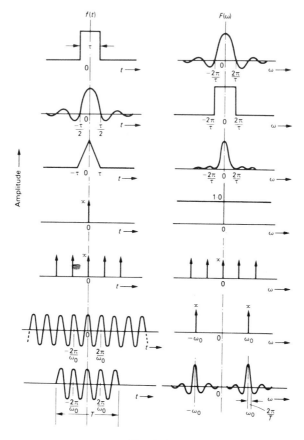

Fig. 2.35

Hence

$$P_{\mathrm{av}} = \sum_{-\infty}^{+\infty} \frac{1}{T} \int_{-T/2}^{+T/2} f(t)\, \mathrm{e}^{jn\omega t}\, \mathrm{d}t$$

$$= \sum_{-\infty}^{+\infty} C_n C_{-n}$$

or

$$P_{\mathrm{av}} = \sum_{-\infty}^{+\infty} |C_n|^2 \ \text{watts}$$

Energy spectrum

For a discontinuous signal, such as a single pulse, the average power tends to zero because $1/T$ tends to zero as T tends to infinity. Hence, it is more meaningful to evaluate the total energy associated with a signal in a resistance of 1 ohm.

Hence

$$\text{energy } W = \int_{-\infty}^{+\infty} f^2(t)\, dt$$

Since

$$f(t) = \frac{1}{2\pi} \int_{-\infty}^{+\infty} F(\omega)\, e^{j\omega t}\, d\omega$$

we obtain

$$W = \int_{-\infty}^{+\infty} f(t)\, \frac{1}{2\pi} \int_{-\infty}^{+\infty} F(\omega)\, e^{j\omega t}\, d\omega\, dt$$

$$= \int_{-\infty}^{+\infty} \frac{d\omega}{2\pi} F(\omega) \int_{-\infty}^{+\infty} f(t)\, e^{j\omega t}\, dt$$

$$= \frac{1}{2\pi} \int_{-\infty}^{+\infty} F(\omega)\, F(-\omega)\, d\omega = \frac{1}{2\pi} \int_{-\infty}^{+\infty} |F(\omega)|^2\, d\omega$$

or

$$W = \int_{-\infty}^{+\infty} |F(f)|^2\, df \text{ joules}$$

This result is known as Rayleigh's theorem and the quantity $|F(f)|^2$ is called the energy spectral density since it is the energy per unit frequency.

2.10 Discrete Fourier transform (DFT)[9]

The Fourier transform can be applied to a continuous signal $y(t)$ which is sampled at discrete intervals of time for a finite sample length. This is shown in Fig. 2.36(a) where the signal $y(t)$ is sampled at intervals of T seconds over N samples giving a sample length of NT. The signal $s(t)$ represents the discrete samples which can be regarded as impulse functions (delta functions) of amplitude $f(nT)$ where $n = 0, 1, 2, \ldots, N-1$ and this is shown in Fig. 2.36(b).

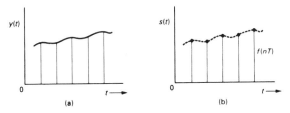

Fig. 2.36

The discrete samples $f(nT)$ can be expressed as

$$s(t) = \sum_{n=0}^{n=N-1} f(nT)\, \delta(t - nT)$$

where $\delta(t - nT)$ is the delta function at $t = nT$.

If such a discrete signal has a spectrum which consists of N frequency components spaced Ω apart, for convenience, we must have

$$N\Omega = 2\pi/T$$

or
$$\Omega = 2\pi/NT$$

Hence, if the Fourier transform of $f(nT)$ is denoted by $F(\omega)$, we have

$$F(\omega) = \int_{-\infty}^{+\infty} s(t) e^{-j\omega t} dt = \int_{-\infty}^{+\infty} \sum_{n=0}^{n=N-1} f(nT) \delta(t - nT) e^{-j\omega t} dt$$

and if the value of ω is assumed to be made up of m multiples of Ω, such that $\omega = m\Omega$, with $m = 0, 1, 2, \ldots, N - 1$, we then obtain

$$F(m\Omega) = \sum_{n=0}^{n=N-1} f(nT) \int_{-\infty}^{+\infty} \delta(t - nT) e^{-j\omega t} dt$$

The expression under the integral sign represents the Fourier transform of an impulse shifted in time by nT and, by the shift theorem, equals $e^{-j\omega nT}$.

Hence
$$F(m\Omega) = \sum_{n=0}^{n=N-1} f(nT) e^{-jmn\Omega T}$$

which is the discrete Fourier transform (DFT) of the N discrete samples denoted by $f(nT)$.

Furthermore, since $e^{-jmn\Omega T} = e^{-j(mn + N)\Omega T}$ where N is an integer, the frequency components repeat themselves after N components. However, as the components after $(N/2 - 1)\Omega$ are mirror images of the preceding components, in practice only about $N/2$ different components need be determined. This is illustrated in Fig. 2.37.

Comments
1.   The inverse discrete Fourier transform is given by the expression

$$f(nT) = \frac{1}{N} \sum_{m=0}^{m=N-1} F(m\Omega) e^{jmn\Omega T}$$

which is easily verified by substituting for $F(m\Omega)$ in the right hand side of the equation, i.e.

$$\text{r.h.s.} = \frac{1}{N} \sum_{m=0}^{m=N-1} f(nT) \sum_{n=0}^{n=N-1} e^{-jmn\Omega T} \times e^{jmn\Omega T}$$

or
$$\text{r.h.s.} = \frac{1}{N} f(nT) N = f(nT)$$

which is the term on the left hand side as given above.

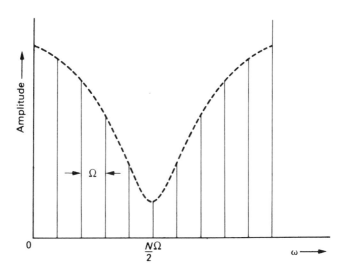

Fig. 2.37

2. Since the z-transform $F(z)$ of $s(t)$ is given by[10]

$$F(z) = \sum_{n=0}^{n=N-1} f(nT) z^{-n}$$

and if $z = e^{jmn\Omega T}$, we obtain

$$F(e^{jmn\Omega T}) = \sum_{n=0}^{n=N-1} f(nT) e^{-jmn\Omega T}$$

which has the same form as the DFT.

Hence, the DFT is a set of numerical values of the z-transform at N equally spaced points around the unit circle since $e^{-jm\Omega T} = e^{-jm2\pi/N}$.

3. By assuming earlier that $N\Omega = 2\pi/T$, i.e. that there are N discrete frequency components, the DFT satisfies the sampling theorem if we have

$$\omega_s = 2\pi/T$$

with

$$\omega_s = 2[(N/2)\Omega]$$

or

$$\omega_s > 2[N/2 - 1]\Omega$$

and so $(N/2 - 1)\Omega$ is the highest, independent frequency component which can be determined.

4.   If the condition $\omega_s > 2(N/2 - 1)\Omega$ is not satisfied, the spectral components overlap near the value of $(N/2)\Omega$. This effect is known as *aliasing* and the spectral components up to $(N/2)\Omega$ cannot be separated out without some distortion.

2.11   Fast Fourier transform (FFT)[9]

The DFT can be written more conveniently by means of the substitution $W = e^{-2\pi j/N}$ which yields

$$e^{-jmn\Omega T} = e^{-j2\pi mn/N} = W^{mn}$$

Hence, the DFT can be written as

$$F(m\Omega) = \sum_{n=0}^{n=N-1} f(nT)W^{mn} \qquad (m = 0, 1, 2, \ldots, N-1)$$

or
$$[F_m] = [W^{mn}][f_n]$$

in matrix form where symbolically we have

$$[F_m] = \begin{bmatrix} F(0) \\ F(\Omega) \\ . \\ . \\ . \\ F(m\Omega) \end{bmatrix}$$

$$[W^{mn}] = \begin{bmatrix} W^0 & W^0 \ldots \ldots \ldots \ldots \ldots & W^0 \\ W^0 & \ldots \ldots \ldots \ldots \ldots \ldots & W^{N-1} \\ . \\ . \\ . \\ W^0 & W^{N-1} \ldots \ldots \ldots \ldots & W^{(N-1)^2} \end{bmatrix}$$

$$[f_n] = \begin{bmatrix} f(0) \\ f(T) \\ . \\ . \\ . \\ f(nT) \end{bmatrix}$$

and 'm' signifies a row while 'n' signifies a column in $[W^{mn}]$.

The evaluation of any frequency component of $F(m\Omega)$, e.g. $F(\Omega)$, involves N complex multiplications and N complex additions, i.e. $2N$ complex values. Hence, the evaluation of $N/2$ *independent* frequency components requires $2N \times (N/2) = N^2$ number of values which for $N = 2^{10} = 1024$ becomes prohibitive.

To reduce considerably the number of operations, various algorithms have been devised. In the well-known Cooley–Tukey algorithm[11], which is given in Appendix C, the number of operations is reduced to $N \log_2 N$ which equals $1024 \log_2 (1024) = 10240$ instead of $(1024)^2$ for $N = 2^{10}$. This yields a ratio of $10240/(1024)^2 \simeq 1 \%$ and amounts to a saving of about 99% in the total number of calculations required.

Example 2.6
For the triangular waveform shown in Fig. 2.38, determine the discrete frequency components using an FFT algorithm for 16 sample points.

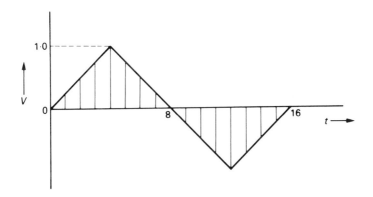

Fig. 2.38

Solution
The discrete values at the sampling points are given in Table 2.1.
Since $N = 16$, we have

$$W = e^{-j2\pi/N} = e^{-j\pi/8} = (0.9239 - j0.3827)$$

$$W^{mn} = (e^{-j\pi/8})^{mn} \quad \text{with} \quad m = 0, 1, 2, \ldots, 15$$

$$n = 0, 1, 2, \ldots, 15$$

$$F(m\Omega) = \sum_{n=0}^{n=15} f(nT)\, W^{mn}$$

Table 2.1

Point	Value	Point	Value
0	0·00	8	0·00
1	0·25	9	− 0·25
2	0·50	10	− 0·50
3	0·75	11	− 0·75
4	1·00	12	− 1·00
5	0·75	13	− 0·75
6	0·50	14	− 0·50
7	0·25	15	− 0·25

Typical calculations for (a) $F(0\Omega)$ and (b) $F(1\Omega)$ are given below. Results for all the spectral components are summarised in Table 2.2.

(a) $F(0\Omega)$
Here, $m = 0$ and $mn = 0$. Hence

$$F(0\Omega) = \sum_{n=0}^{n=5} f(nT)\, W^0$$

with

$$W^0 = (e^{-j\pi/8})^0 = e^0 = 1$$

and

$$F(0\Omega) = f(0) + f(T) + \ldots + f(15T)$$

Substituting for the sample values from Table 2.1 yields

$$F(0\Omega) = 0·00 + 0·25 + 0·50 + 0·75 + 1·00 + 0·75 + 0·50 + 0·25$$
$$+ 0·00 - 0·25 - 0·50 - 0·75 - 1·00 - 0·75 - 0·50 - 0·25$$

or

$$F(0\Omega) = 0$$

(b) $F(1\Omega)$
Here, $m = 1$ and so $mn = n$. Hence

$$F(1\Omega) = \sum_{n=0}^{n=15} f(nT)\, W^n$$

with

$$W^0 = 0$$
$$W^1 = 0·9239 - j\,0·3827$$
$$W^2 = 0·7070 - j\,0·7070$$
$$W^3 = 0·3827 - j\,0·9239$$
$$W^4 = -j$$
$$W^5 = -(0·3827 + j\,0·9239)$$
$$W^6 = -(0·7070 + j\,0·7070)$$
$$W^7 = -(0·9239 + j\,0·3827)$$
$$W^8 = 0$$

The remaining values W^9 to W^{15} are complex conjugates of the previous values, e.g. $W^9 = (-0.9239 + j\, 0.3827)$ which is the complex conjugate of W^7. Hence

$$F(1\Omega) = f(0)\, W^0 + f(T)\, W^1 + f(2T)\, W^2 + \ldots + f(15T)\, W^{15}$$

and substituting for the sample values from Table 2.2 yields

$$\begin{aligned}
F(1\Omega) = \; & 0.00 + 0.25(0.9239 - j\, 0.3827) \\
& + 0.50(0.7070 - j\, 0.7070) \\
& + 0.75(0.3827 - j\, 0.9239) \\
& + 1.00(-j) + 0.75(-0.3827 - j\, 0.9239) \\
& + 0.50(-0.7070 - j\, 0.7070) \\
& + 0.25(-0.9239 - j\, 0.3827) \\
& + 0.00 - 0.25(-0.9239 + j\, 0.3827) \\
& - 0.50(-0.7070 + j\, 0.7070) \\
& - 0.75(-0.3827 + j\, 0.9239) \\
& - 1.00(j) - 0.75(0.3827 + j\, 0.9239) \\
& - 0.50(0.7070 + j\, 0.7070) \\
& - 0.25(0.9239 + j\, 0.3827)
\end{aligned}$$

or
$$F(1\Omega) = 1.7429 - 1.7429 - j\, 6.5684$$

Hence
$$F(1\Omega) = -j\, 6.5684$$

or
$$|F(1\Omega)| = 6.5684$$

Table 2.2

F_m	$|F_m|$
F_0	0
$F_1,\ F_{15}$	6.5684
$F_2,\ F_{14}$	0
$F_3,\ F_{13}$	0.8099
$F_4,\ F_{12}$	0
$F_5,\ F_{11}$	0.3616
$F_6,\ F_{10}$	0
$F_7,\ F_9$	0.2599
F_8	0

Comments
1. The real part of F_m in all cases is zero.
2. The values of F_9 to F_{15} are the complex conjugates of F_7 to F_1 respectively.

3

Network response

3.1 Non-repetitive waveforms

From a communication point of view, a knowledge of the frequency spectrum of repetitive and non-repetitive waveforms can help to show how the signal will be affected when passing through a network with a frequency characteristic of its own. The overall result is called the network response.

The relationship between the input *frequency* function and the response is purely algebraic and easier to obtain than the alternative relation between the input *time* function and the response. The Fourier transform technique will now be used to obtain the response for a non-repetitive waveform or pulse.

Let $H(\omega)$ be the transfer function of the network shown in Fig. 3.1, which is generally complex and defined by the equation

$$\text{transfer function} = \frac{\text{output transform}}{\text{input transform}} = H(\omega) = |H(\omega)|\underline{/\phi(\omega)}$$

where $|H(\omega)|$ is the amplitude response and $\phi(\omega)$ is the phase shift through the network. Both quantities are dependent on ω.

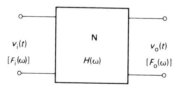

Fig. 3.1

If $v_i(t)$ and $v_o(t)$ are the input and output time functions respectively, and $F_i(\omega)$ and $F_o(\omega)$ are the corresponding Fourier transforms, we have

$$\frac{F_o(\omega)}{F_i(\omega)} = H(\omega) \quad \text{if the system is linear}$$

or

$$F_o(\omega) = H(\omega)\, F_i(\omega)$$

where

$$F_i(\omega) = \int_{-\infty}^{+\infty} v_i(t)\, e^{-j\omega t} dt$$

The output $v_o(t)$ is then obtained from the inverse Fourier transform which gives

$$v_o(t) = \frac{1}{2\pi} \int_{-\infty}^{+\infty} F_o(\omega)\,e^{j\omega t}\,d\omega$$

or

$$v_o(t) = \frac{1}{2\pi} \int_{-\infty}^{+\infty} H(\omega)\,F_i(\omega)\,e^{j\omega t}\,d\omega$$

Notes
1. If $v_i(t)$ is a d.c. or single sinusoidal voltage, $H(\omega)$ is also equal to the ratio of the output voltage to the input voltage.
2. The output voltage can also be found by means of the convolution integral (see Section 3.6).

3.2 Ideal low-pass filter

The most commonly used networks are called filters and a typical example is the low-pass filter. For the purpose of this analysis, it is convenient to consider an ideal low-pass filter which has the characteristics shown in Fig. 3.2. Positive and negative frequencies are considered as a mathematical generality.

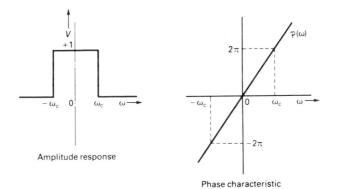

Amplitude response

Phase characteristic

Fig. 3.2

The filter is characterised by the following:
1. A constant amplitude response from $-\omega_c$ to $+\omega_c$.
2. A sharp vertical cut-off at $-\omega_c$ and $+\omega_c$.
3. A linear phase delay.

Hence

$$H(\omega) = 1 \times e^{-j\phi(\omega)}$$

It is convenient now to examine the behaviour of such an ideal filter on well-known time functions and obtain useful results which may be applied to more practical filters.

(a) Step-function response

The unit step function shown in Fig. 3.3 can be considered as the sum of two waveforms. The first is simply a d.c. voltage of magnitude $\frac{1}{2}$ applied to the system at *all time* and the other is a *unit step* voltage from $-\frac{1}{2}$ to $+\frac{1}{2}$, applied at $t = 0$.

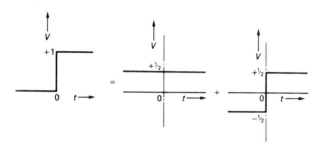

Fig. 3.3

Let the output responses to these waveforms be $v'_o(t)$ and $v''_o(t)$ respectively. The output response of the first waveform is obtained from the relationship

$$\frac{v'_o(t)}{v_i(t)} = H(\omega) = 1 \times e^{-j\phi(\omega)} \quad \text{since } v_i(t) \text{ is a d.c. voltage}$$

Now $\phi(\omega) = 0$ at $\omega = 0$ and $v_i(t) = \frac{1}{2}$.

Hence
$$v'_o(t) = \tfrac{1}{2} e^0 = \tfrac{1}{2}$$

The response $v''_o(t)$ to the unit step voltage from $-\frac{1}{2}$ to $+\frac{1}{2}$ is found by first obtaining its Fourier transform $F_i(\omega)$ as in Section 2.7(d) and then using the inverse Fourier transform to obtain $v''_o(t)$.

Hence, we have
$$F_i(\omega) = \frac{1}{j\omega}$$

$$v''_o(t) = \frac{1}{2\pi} \int_{-\infty}^{+\infty} F_o(\omega) e^{j\omega t} d\omega = \frac{1}{2\pi} \int_{-\infty}^{+\infty} H(\omega) F_i(\omega) e^{j\omega t} d\omega$$

where $H(\omega) = 1 \times e^{-j\phi(\omega)}$ and $-\omega_c < \omega < +\omega_c$.

Hence
$$v_0''(t) = \frac{1}{2\pi} \int_{-\omega_c}^{+\omega_c} 1 \times e^{-j\phi(\omega)} \frac{1}{j\omega} e^{j\omega t} \, d\omega$$

$$= \frac{1}{\pi j\omega} \int_0^{\omega_c} e^{j\omega t} e^{-j\phi(\omega)} \, d\omega$$

or
$$v_0''(t) = \frac{1}{\pi j\omega} \int_0^{\omega_c} e^{j\omega(t - t_d)} \, d\omega \quad \text{where} \quad \omega t_d = \phi(\omega)$$

Hence
$$v_0''(t) = \frac{1}{\pi} \int_0^{\omega_c} \frac{\sin \omega(t - t_d)}{\omega} \, d\omega$$

by using the sine term only because the *second* step function (see Fig. 3.3) is an odd function of time.

To simplify the integral, put $\omega(t - t_d) = x$ and $\omega_c(t - t_d) = X$. Hence, $d\omega = dx/(t - t_d)$ and we obtain

$$v_0''(t) = \frac{1}{\pi} \int_0^X \frac{\sin x}{x} \, dx$$

This is called the 'sine-integral' of x and is obtained by integrating the area under the $(\sin x)/x$ curve up to some value X. It is shown in Fig. 3.4.

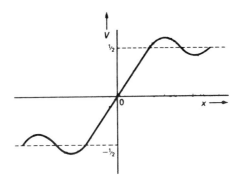

Fig. 3.4

The total output response $v_0(t)$ of the network is then given by

$$v_0(t) = v_0'(t) + v_0''(t)$$

or
$$v_0(t) = \frac{1}{2} + \frac{1}{\pi} \int_0^X \frac{\sin x}{x} \, dx$$

which is shown in Fig. 3.5.

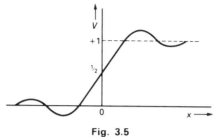

Fig. 3.5

Comments

1. Defining the rise time t_r as from 0·1 to 0·9 of the amplitude gives $\omega_c t_r = \pi$

$$\therefore \quad t_r = 1/2f_c$$

2. The ripple frequency f_r is given by $f_r = 1/T$ where $\omega_c T = 2\pi$

$$\therefore \quad f_r = f_c$$

3. The time delay to the half height is t_d where $\omega t_d = \phi(\omega)$

$$t_d = \frac{\phi(\omega)}{\omega} \simeq \frac{2\pi}{\omega_c}$$

or $$t_d \simeq 1/f_c$$

4. The output response appears to begin even before the application of the step function. This is because of the idealised filter characteristics which cannot be obtained in practice. Hence, the response in a practical filter will occur *entirely* after the step has been applied.

(b)   Rectangular pulse
If the width of the pulse $\tau \gg 1/f_c$, the response is that due to two step functions in opposite phase and delayed by an amount equal to τ. This is illustrated below in Fig. 3.6.

(c)   Impulse function (unit impulse)
Treat this as a very narrow rectangular pulse whose width $\tau < 1/f_c$. The response has the familiar $(\sin x)/x$ shape.
If $v_o(t)$ is the output response, then

$$v_o(t) = \frac{1}{2\pi} \int_{-\infty}^{+\infty} H(\omega) F_i(\omega) e^{j\omega t} d\omega$$

where $H(\omega) = 1 \times e^{-j\phi(\omega)}$ and $F_i(\omega) = A\tau = 1$ (Section 2.7(c)).

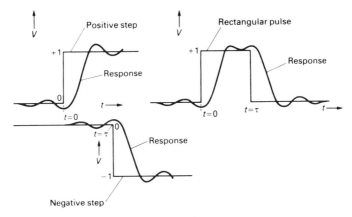

Fig. 3.6

Hence
$$v_o(t) = \frac{1}{2\pi} \int_{-\omega_c}^{+\omega_c} e^{-j\phi(\omega)} e^{j\omega t} d\omega$$

$$= \frac{1}{2\pi} \int_{-\omega_c}^{+\omega_c} e^{j\omega(t-t_d)} d\omega \quad \text{where} \quad \omega t_d = \phi(\omega)$$

$$= \frac{1}{2\pi} \left[\frac{e^{j\omega(t-t_d)}}{j(t-t_d)} \right]_{-\omega_c}^{+\omega_c}$$

or
$$v_o(t) = \frac{1}{2\pi j} \left[\frac{e^{j\omega_c(t-t_d)} - e^{-j\omega_c(t-t_d)}}{(t-t_d)} \right]$$

Hence
$$v_o(t) = \frac{\omega_c}{\pi} \frac{\sin \omega_c(t-t_d)}{\omega_c(t-t_d)} = \frac{\omega_c}{\pi} \frac{\sin x}{x}$$

where $x = \omega_c(t-t_d)$. The response is shown in Fig. 3.7 for $A\tau = 1$, i.e. $A = 1/\tau$. Hence, a pulse of unit amplitude would have a peak response of $\omega_c \tau/\pi$.

3.3 Repetitive waveforms

The response of the filter is obtained by first determining the frequency components of the input waveform by the usual Fourier series technique. Of these, only those which the filter will pass, will appear in the output. From a knowledge of the cut-off frequency f_c (low- or high-pass filter) or the two cut-off frequencies f_{c1} and f_{c2} (band-pass or band-stop filter), it remains to determine by inspection which of the input frequencies will appear at the output.

The simple superposition of the output frequencies will give the output response of the filter (see Examples 2.2 and 2.3).

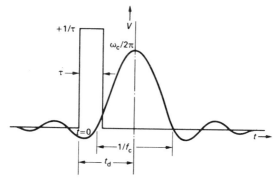

Fig. 3.7

3.4 Practical low-pass filter

The transfer function has amplitude and phase characteristics as shown in Fig. 3.8.

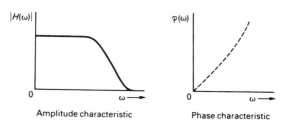

Amplitude characteristic   Phase characteristic

Fig. 3.8

An alternative way of expressing the amplitude response is in terms of the 'attenuation' of the network. The magnitude of the signal is reduced when it passes through the network and the attenuation loss is usually given in decibels (dB).

Calculation
For single sinusoidal voltages V_o and V_i, we have

$$\frac{V_o}{V_i} = H(\omega) = |H(\omega)| \, e^{-j\phi(\omega)}$$

Hence
$$\left|\frac{V_o}{V_i}\right| = |H(\omega)|$$

The attenuation loss in decibels is defined as

$$\text{attenuation loss} = 20 \log_{10} \left| \frac{V_i}{V_o} \right| \, \text{dB}$$

or $\qquad \text{attenuation loss} = 20 \log_{10} \frac{1}{|H(\omega)|} \, \text{dB}$

and the phase shift through the network is $\phi(\omega)$ radians or degrees.

Example 3.1

A time-varying voltage $f(t)$ with t measured in microseconds and defined by

$$f(t) = 0 \qquad \text{for} \quad -\infty \leqslant t \leqslant 0$$
$$f(t) = 2e^{-2t} \qquad \text{for} \quad 0 \leqslant t \leqslant \infty$$

is applied to the input of a linear, passive, two-terminal-pair network. The corresponding voltage $g(t)$, at the output, is defined by

$$g(t) = 0 \qquad\qquad\qquad\qquad \text{for} \quad -\infty \leqslant t \leqslant 0$$
$$g(t) = \tfrac{2}{3}e^{-2t} - 2e^{-t} + \tfrac{4}{3}e^{-t/2} \quad \text{for} \quad 0 \leqslant t \leqslant \infty$$

Using a Fourier integral, derive expressions for the spectral function of the input and output voltages and, hence, for the voltage transfer ratio of the network. Calculate
(a) the attenuation in dB at zero frequency,
(b) the frequency at which the attenuation is 3 dB greater than the zero-frequency value. (U.L.)

Solution
Input Let $F_i(\omega)$ be its Fourier transform (Fourier integral).

Hence $\qquad F_i(\omega) = \displaystyle\int_{-\infty}^{+\infty} 2e^{-2t} \, e^{-j\omega t} \, dt = \int_0^\infty 2e^{-(2+j\omega)t} \, dt$

$$= 2 \left[\frac{e^{-(2+j\omega)t}}{-(2+j\omega)} \right]_0^\infty = \frac{2}{2+j\omega}$$

Output Let $F_o(\omega)$ be the Fourier transform.

Hence $\quad F_o(\omega) = \displaystyle\int_{-\infty}^{+\infty} \left[\tfrac{2}{3}e^{-(2+j\omega)t} - 2e^{-(1+j\omega)t} + \tfrac{4}{3}e^{-[(1/2)+j\omega]t} \right] dt$

$$= \int_0^\infty \tfrac{2}{3}e^{-(2+j\omega)t} \, dt - \int_0^\infty 2e^{-(1+j\omega)t} \, dt + \int_0^\infty \tfrac{4}{3}e^{-[(1/2)+j\omega]t} \, dt$$

or $\qquad F_o(\omega) = \dfrac{2}{3(2+j\omega)} - \dfrac{2}{(1+j\omega)} + \dfrac{4}{3(\frac{1}{2}+j\omega)}$

Now $\qquad H(\omega) = \dfrac{F_o(\omega)}{F_i(\omega)} = \dfrac{1}{3} - \dfrac{2+j\omega}{1+j\omega} + \dfrac{2(2+j\omega)}{3(\frac{1}{2}+j\omega)}$

or
$$H(\omega) = \frac{1}{(1 + 2j\omega)(1 + j\omega)} \quad \text{and} \quad H(0) = 1$$

(a)   Attenuation $= 20 \log_{10} |V_i/V_o| = 20 \log_{10} 1/|H(0)| = 20 \log_{10} 1 = 0 \, \text{dB}$.
(b)   For 3 dB point

$$|1 + 2j\omega| \, |1 + j\omega| = \sqrt{2} \quad \text{i.e.} \quad H(\omega) = \frac{H(0)}{\sqrt{2}} = \frac{1}{\sqrt{2}}$$

or
$$(1 + 4\omega^2)(1 + \omega^2) = 2$$

Put
$$x = \omega^2 \quad \text{or} \quad (4x + 1)(x + 1) = 2$$

Hence
$$x = 0.175 \quad \text{(using the positive value only)}$$

or
$$\omega = 0.4175$$

and
$$f = 66.5 \, \text{kHz}$$

3.5   The Laplace transform[12]

In many engineering problems, the functions of main interest are those which begin at $t = 0$. In this case, it is easier to employ the Laplace transform technique. This also applies in cases when the function does not lead to a finite solution by the use of the Fourier transform and can be made to do so by the Laplace transform.

The essential difference between the two methods is simply the fact that the Fourier transform employs a summation of waves of positive and negative frequencies, while the Laplace transform uses *damped* waves through the use of an additional factor $e^{-\sigma}$ where σ is a positive number.

Definition
For time functions $f(t)$, either periodic or non-periodic, which are defined from $t = 0$ and thereafter, the Laplace transform $\mathscr{L}[f(t)]$ is given by

$$\mathscr{L}[f(t)] = \int_0^\infty f(t) \, e^{-(\sigma + j\omega)t} \, dt$$

Putting $s = \sigma + j\omega$ and $\mathscr{L}[f(t)] = F(s)$, for convenience, yields

$$F(s) = \int_0^\infty f(t) \, e^{-st} \, dt$$

Example 3.2
Obtain the Laplace transform of the unit step-function waveform of Section 2.7(d).

Solution

$$F(s) = \int_0^\infty f(t)\, e^{-st}\, dt$$

$$= \int_0^\infty 1 \times e^{-st}\, dt \quad \text{because} \quad f(t) = 1 \quad \text{for} \quad t > 0$$

$$= \frac{[e^{-st}]_0^\infty}{-s} = \frac{e^{-\infty} - e^0}{-s} = \frac{1}{s}$$

or

$$F(s) = \frac{1}{\sigma + j\omega}$$

Applications

The function $F(s)$ involves only the parameter s and is an algebraic expression. Because of this simplification and the use of standard rules, it is easier to convert the time function $f(t)$ to the new function $F(s)$ when solving problems. The problem is then solved in the complex s-*domain* and finally converted back to the *time domain* by using the inverse Laplace transform obtained from standard tables.

This is particularly true of differential equations which are easier to manipulate in the s-domain and can then be converted back to the time domain for the final physical solution.

To assist in the application of the Laplace transform to differential equations, it is necessary to be acquainted with the basic results for various standard time functions and these are given in Table 3.1.

Differential equations

The equation in the time domain is converted to the s-domain by obtaining the Laplace transform of each term in the equation. For example, assume $i(t) = f(t)$, $i(0) = f(0)$, and $i_n(0) = d^n i(t)/dt^n$ (at $t = 0$). Standard expressions for this case are given in Table 3.2.

Inverse Laplace transform

To obtain $f(t)$ from $F(s)$, the inverse Laplace transform is required. This is mathematically expressed as a complex integral, where integration is along a straight line in the s-plane at $s = \sigma_1$, from $s = \sigma_1 - j\infty$ to $s = \sigma_1 + j\infty$.

We have

$$f(t) = \mathscr{L}^{-1}[F(s)] = \frac{1}{2\pi j} \int_{\sigma_1 - j\infty}^{\sigma_1 + j\infty} F(s)\, e^{st}\, ds$$

In practice, it is not usual to carry out this cumbersome integration. To obtain the solution of a differential equation, it is only necessary to arrange the final expression obtained in the s-domain in a standard form. This is done by

Table 3.1

$f(t)$	$F(s)$
a (constant)	a/s
e^{-at}	$1/(s+a)$
$1-e^{-at}$	$a/s(s+a)$
t^n (n an integer)	n/s^{n+1}
te^{-at}	$1/(s+a)^2$
$\sin \omega t$	$\omega/(s^2+\omega^2)$
$\cos \omega t$	$s/(s^2+\omega^2)$
$\dfrac{t}{2\omega}\sin \omega t$	$\dfrac{s}{(s^2+\omega^2)}$
$\dfrac{t}{2\omega}\cos \omega t$	$\dfrac{(s^2-\omega^2)}{(s^2+\omega^2)}$
$\sin(\omega t+\phi)$	$\dfrac{s\sin \phi +\omega\cos \phi}{(s^2+\omega^2)}$
$e^{-at}\sin \omega t$	$\omega/\left[(s+a)^2+\omega^2\right]$
$e^{-at}\cos \omega t$	$(s+a)/\left[(s+a)^2+\omega^2\right]$
$\sinh \omega t$	$\omega/(s^2-\omega^2)$
$\cosh \omega t$	$s/(s^2-\omega^2)$

Table 3.2

$f(t)$	$F(s)$
$i(t)$	$I(s)$
$ai(t)$ (a is constant)	$aI(s)$
$\dfrac{di(t)}{dt}$	$sI(s)-i(0)$
$\dfrac{d^n i(t)}{dt^n}$	$s^n I(s)-s^{n-1}i(0)-s^{n-2}i_1(0)$ $- \ldots s^0 i_{n-1}(0)$
$\displaystyle\int i(t)\,dt$	$\dfrac{I(s)}{s}$

resolving the function $F(s)$ into partial fractions according to standard rules. It involves the use of constants, A, B, etc., which are evaluated either by equating coefficients on either side or by the substitution of appropriate values for s which make certain terms zero.

Standard rules

1. Real roots s_1, s_2, etc.

    Put $F(s) = \dfrac{A}{(s+s_1)} + \dfrac{B}{(s+s_2)} + \cdots$

2. Real *repeated* roots s_1, s_1, etc.

    Put $F(s) = \dfrac{A}{(s+s_1)} + \dfrac{B}{(s+s_1)^2} + \dfrac{C}{(s+s_1)^3}$ (for three repeated roots)

3. Imaginary roots $\alpha + j\omega$, $\alpha - j\omega$.

    Put $F(s) = \dfrac{A}{(s+\alpha)+j\omega} + \dfrac{B}{(s+\alpha)-j\omega}$

The time function $f(t)$ is then obtained by identifying each partial fraction with an inverse Laplace transform from Table 3.1, i.e. by using column $F(s)$ to obtain the corresponding $f(t)$. This is illustrated in Examples 3.3 and 3.4.

Example 3.3

Two circuits are coupled by mutual inductance M. A switch S is closed at time $t = 0$ and a voltage E_0 is applied to the primary. If the currents i_1 and i_2 are zero at $t = 0$, find an expression for the secondary current at time t. The circuit arrangement is given in Fig. 3.9.

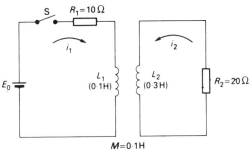

Fig. 3.9

Solution

Assuming the current directions are as shown for positive mutual, the circuit equations are

$$L_1 \frac{di_1}{dt} + R_1 i_1 + M \frac{di_2}{dt} = E_0$$

$$L_2 \frac{di_2}{dt} + R_2 i_2 + M \frac{di_1}{dt} = 0$$

Taking transforms for both sides gives

$$L_1[sI_1(s) - i_1(0)] + R_1 I_1(s) + M[sI_2(s) - i_2(0)] = \frac{E_0}{s}$$

$$L_2[sI_2(s) - i_2(0)] + R_2 I_2(s) + M[sI_1(s) - i_1(0)] = 0$$

Since $i_1(0) = i_2(0) = 0$ at $t = 0$, we have

$$(L_1 s + R_1)I_1(s) + MsI_2(s) = \frac{E_0}{s}$$

$$(L_2 s + R_2)I_2(s) + MsI_1(s) = 0$$

Eliminating $I_1(s)$ between these equations gives

$$I_2(s) = \frac{ME_0}{(M^2 - L_1 L_2)s^2 - (L_1 R_2 + L_2 R_1)s - R_1 R_2}$$

Substituting the numerical values yields

$$M^2 - L_1 L_2 = -0.05$$

$$L_1 R_2 + L_2 R_1 = 7.0$$

$$R_1 R_2 = 200$$

or $$I_2(s) = -\frac{0.1 E_0}{(0.05 s^2 + 7s + 200)} = -\frac{2E_0}{(s^2 + 140s + 4000)}$$

By the method of partial fractions, we obtain

$$I_2(s) = -2E_0 \left\{ \left[\frac{A}{(s + 100)} \right] + \left[\frac{B}{(s + 40)} \right] \right\}$$

and $$A(s + 40) + B(s + 100) = 1$$

giving $$A + B = 0 \quad \text{or} \quad A = -B$$

Hence $$40A + 100B = 1 \quad \text{or} \quad B = \tfrac{1}{60} \quad \text{and} \quad A = -\tfrac{1}{60}$$

with $$I_2(s) = \frac{E_0}{30} \left[\frac{1}{(s + 100)} - \frac{1}{(s + 40)} \right]$$

Taking inverse transforms yields

$$i_2 = \frac{E_0}{30} \left[e^{-100t} - e^{-40t} \right]$$

Example 3.4

A coil of inductance L and resistance R is connected in parallel with a capitance C. At $t = 0$, a constant current E/R is fed into the parallel combination when the initial charge and current are zero. Assuming that $R^2/4L^2 < 1/LC$, find the voltage across the capacitor at time t. The circuit is shown in Fig. 3.10.

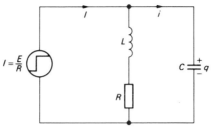

Fig. 3.10

Solution
Let the voltage across the capacitor be V when its charging current is i. We have

$$q/C = V$$

$$L\frac{\mathrm{d}}{\mathrm{d}t}(I - i) + R(I - i) = V$$

or
$$\frac{q}{C} = L\frac{\mathrm{d}}{\mathrm{d}t}(I - i) + R(I - i)$$

Taking transforms on both sides yields

$$\frac{Q(s)}{C} = \mathscr{L}\left[L\frac{\mathrm{d}}{\mathrm{d}t}(I - i) + R(I - i)\right]$$

Now
$$\mathscr{L}\left[L\frac{\mathrm{d}}{\mathrm{d}t}(I - i)\right] = Ls\left[\frac{E}{Rs} - I(s)\right] \qquad \{(I - i) = 0 \text{ at } t = 0\}$$

and
$$\mathscr{L}\left[R(I - i)\right] = R\left[I/s - I(s)\right] \qquad \{(I - i) = 0 \text{ at } t = 0\}$$

Since
$$i = \frac{\mathrm{d}q}{\mathrm{d}t}$$

hence
$$I(s) = sQ(s) - q(0) = sQ(s) \qquad \{q(0) = 0 \text{ at } t = 0\}$$

and
$$\frac{Q(s)}{C} = \frac{EL}{R} - Ls^2 Q(s) + \frac{E}{s} - RsQ(s)$$

or
$$\left(Ls^2 + Rs + \frac{1}{C}\right)Q(s) = \frac{EL}{R} + \frac{E}{s}$$

and
$$Q(s) = \frac{EL/R + E/s}{(Ls^2 + Rs + 1/C)} = \frac{E/R + E/Ls}{(s^2 + R/Ls + 1/LC)}$$

Let
$$\alpha = R/2L \quad \text{and} \quad \beta^2 = 1/LC - R^2/4L^2$$

or
$$\alpha^2 + \beta^2 = 1/LC$$

Hence
$$Q(s) = \frac{E}{R}\left[\frac{1}{(s + \alpha)^2 + \beta^2}\right] + \frac{E}{L}\left[\frac{A}{s} + \frac{(Bs + D)}{(s + \alpha)^2 + \beta^2}\right]$$

where A, B, and D are constants such that

$$A[(s+\alpha)^2 + \beta^2] + (Bs + D)s \equiv 1$$

or

$$A + B = 0$$

$$2\alpha A + D = 0$$

$$A(\alpha^2 + \beta^2) = 1$$

giving

$$A = \frac{1}{(\alpha^2 + \beta^2)} = LC$$

$$B = -\frac{1}{(\alpha^2 + \beta^2)} = -LC$$

$$D = -\frac{2\alpha}{\alpha^2 + \beta^2} = -2\alpha LC$$

Hence

$$Q(s) = \frac{E}{R}\left[\frac{1}{(s+\alpha)^2 + \beta^2}\right] + \frac{E}{L}\left[\frac{LC}{s} - LC\left\{\frac{(s+\alpha)}{(s+\alpha)^2 + \beta^2} + \frac{\alpha}{(s+\alpha)^2 + \beta^2}\right\}\right]$$

or

$$\frac{Q(s)}{C} = \frac{E}{RC}\left[\frac{1}{(s+\alpha)^2 + \beta^2}\right] + \frac{E}{s} - E\left\{\frac{(s+\alpha)}{(s+\alpha)^2 + \beta^2} + \frac{\alpha}{(s+\alpha)^2 + \beta^2}\right\}$$

Taking inverse transforms yields

$$\frac{q}{C} = E\left[1 - e^{-\alpha t}\left\{\cos\beta t + \left(\frac{R}{2\beta L} - \frac{1}{\beta RC}\right)\sin\beta t\right\}\right]$$

which is the voltage across the capacitor C.

3.6   Convolution integral

The output response of a linear network in the time domain may be obtained by the process of *convolution*.

The convolution of two functions $u(t)$ and $v(t)$ is defined by the convolution integral

$$u(t) * v(t) = \int_{-\infty}^{+\infty} u(\tau)\, v(t - \tau)\, \mathrm{d}\tau = \int_{-\infty}^{+\infty} v(\tau)\, u(t - \tau)\, \mathrm{d}\tau$$

where the asterisk denotes convolution and τ is an arbitrary variable which symbolises *excitation* time, while the real variable t symbolises *response* time.

The significance of convolution is shown graphically in Fig. 3.11 where $u(\tau)$ and $v(\tau)$ are arbitrary functions of the dummy variable τ. To obtain $v(t - \tau)$, we first obtain $v(-\tau)$ and move the waveform forward by time t to obtain $v(t - \tau)$. The product $u(\tau)\, v(t - \tau)$ is then obtained for overlapping values of τ and the area under the graph $u(\tau)\, v(t - \tau)$ represents the convolution integral $u(t) * v(t)$. Convolution therefore involves convolving or turning over one of the waveforms, such as $v(t)$.

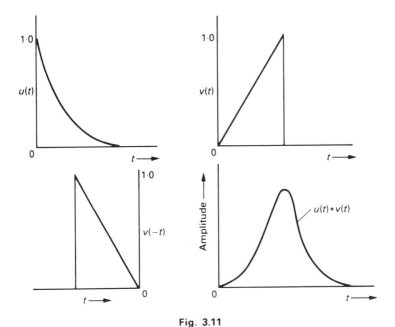

Fig. 3.11

The convolution integral can be used to obtain the output response $v_o(t)$ of a linear network for a given input signal $v_i(t)$ if the impulse response $h(t)$ of the network to an input impulse $\delta(t)$ is known. In this case, if $u(t) = h(t)$ and $v(t) = v_i(t)$, we obtain

$$v_o(t) = h(t) * v_i(t)$$

or $\qquad v_o(t) = \int_{-\infty}^{+\infty} h(\tau)\,v_i(t-\tau)\,d\tau = \int_{-\infty}^{+\infty} v_i(\tau)\,h(t-\tau)\,d\tau$

and the output signal is thus obtained by convolving the input signal with the impulse response of the network.

Comment
It can be shown that convolution in the time domain is equivalent to multiplication in the frequency domain and vice versa (see Section 3.1).

Example 3.5
A unit step voltage is applied to the *RC* integrating network shown in Fig. 3.12(a). Obtain the output response of the network.

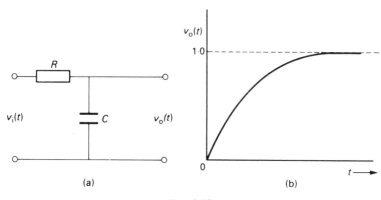

Fig. 3.12

Solution

To obtain the output voltage by the convolution theorem, we require the impulse response $h(t)$ of the network. With the aid of the Laplace transform, we obtain

$$\frac{v_o(s)}{v_i(s)} = H(s)$$

where $H(s)$ is the transfer function of the network with $s = j\omega$. In addition, if $RC = \alpha$, we have

$$H(s) = \frac{1}{(1 + s\alpha)} = \frac{(1/\alpha)}{(s + 1/\alpha)}$$

For an impulse function $\delta(t)$, $v_i(s) = 1$ and we obtain

$$v_o(s) = H(s)\, v_i(s) = H(s)$$

or

$$v_o(s) = \frac{(1/\alpha)}{(s + 1/\alpha)}$$

and from tables the inverse transform is

$$h(t) = \frac{e^{-t/\alpha}}{\alpha}$$

From the convolution integral, we have

$$v_o(t) = \int_{-\infty}^{+\infty} h(t)\, v_i(t - \tau)\, d\tau$$

Hence

$$v_o(t) = \int_0^t \frac{e^{-t/\alpha}}{\alpha}\, d\tau = -[e^{-t/\alpha}]_0^t$$

or

$$v_o(t) = 1 - e^{-t/\alpha} = 1 - e^{-t/RC}$$

which is illustrated in Fig. 3.12(b).

4
Signal transmission

Signals are transmitted through networks, lines, or free space and, in order to make the most economical use of the bandwidth used, multichannel operation[13, 14] is generally employed. This means that several messages or channels are sent simultaneously through the medium. There are three ways of doing this – (a) on a frequency division basis, (b) on a time division basis, and (c) on a code division basis.

4.1 Frequency division basis

This is by far the most commonly used type of transmission and is illustrated in Fig. 4.1. Each message or channel is allocated a band of frequencies and the various frequency bands lie side by side, making a total of around 24 in a telegraph system or as many as 10 800 in a telephone coaxial system. The frequencies are all mixed together at the transmitting end and sent along the same communication path. The system is known as frequency division multiplex (FDM), which has been made possible because of the existence of the band-pass filter. The frequency band of each channel can be separated from the rest of the frequencies by the use of appropriate filters at the receiving end and so the various messages can be received separately and correctly.

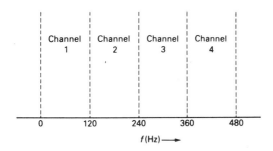

Fig. 4.1

Figure 4.2 shows a typical FDM arrangement for a voice-frequency telegraph system. Each input message modulates a different carrier and the modulated

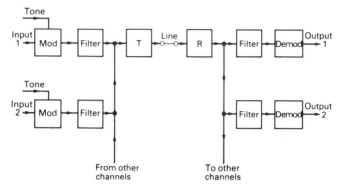

Fig. 4.2

carrier is band-limited by a transmit filter. The various channels are then combined and passed down a common cable or line. At the receiving end, the various channels are separated by receive filters and then demodulated.

Either amplitude modulation or frequency modulation may be used. The main problem is crosstalk caused by the overlap of frequencies in adjacent bands, due to filter characteristics or the presence of intermodulation products arising from circuit non-linearities.

4.2   Time division basis

An alternative method which achieves similar results is to allocate certain intervals of time to each message or channel, each interval being delayed by a small amount and thereby retaining its identity.

This means that basically a form of pulse system is employed and the overall bandwidth is the same as that for each channel; economy of bandwidth is obtained by interleaving the pulses. This is shown in Fig. 4.3 and the system is called time division multiplex (TDM).

Because of the time separation between the channels, each channel can be properly separated at the receiving end. The various messages are recovered correctly by passing the separate pulse trains through separate low-pass filters.

Figure 4.4 shows a schematic arrangement for operating a TDM system. The various input signals are band-limited by filters and selected in time sequence by an electronic 'switch', i.e. a time-gate circuit. The sequential signals are then transmitted over the line or cable to the other end.

At the receiving end, the receiver's signals are selected by another electronic switch which is synchronised with that at the transmitter. The various channels are thereby separated and passed through separate low-pass filters to recover the original messages.

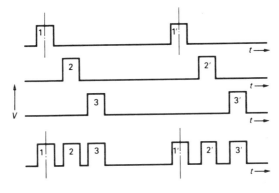

Fig. 4.3

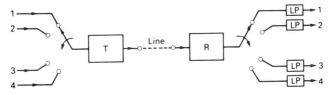

Fig. 4.4

Various types of modulation may be employed, such as pulse amplitude modulation (PAM), pulse time modulation (PTM), and pulse code modulation (PCM). They have certain advantages over one another, but increase in complexity from PAM through PTM to PCM. Crosstalk reduction is possible, but at the expense of increased bandwidth, as compared with FDM.

4.3 Code division basis

In the third form of multiplexing, which corresponds to a spread-spectrum technique,[15] all channels share a common bandwidth and are transmitted at the same time. Each channel is allocated a distinct waveform pattern by modulation with a unique code sequence and it is known as code division multiplexing (CDM).

The code sequences used may belong to an orthogonal group code, such as the set of Walsh functions which are described in Appendix G, or they may be discrete shifts of a pseudo-random code.[16] Demultiplexing at the receiver is based on a pattern-recognition process using correlation detection. The orthogonality requirements met with in both frequency and time division multiplexing correspond here to conditions of zero correlation between channel patterns.

Consider the case of the multiplexing of two baseband digital signals $s_1(t)$ and $s_2(t)$. The carriers used are two pseudo-random code sequences $c_1(t)$ and $c_2(t)$ which are shown in Fig. 4.5. The baseband signals are multiplied with their respective coded carriers and linearly added prior to transmission over a transmission line, as illustrated in Fig. 4.6. At the receiver, cross-correlation with stored replicas of the carrier sequences $c_1(t)$ and $c_2(t)$ yields the corresponding baseband signals $s_1(t)$ and $s_2(t)$. For optimum detection, *matched filters** are

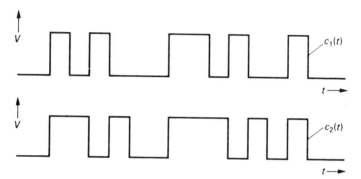

Fig. 4.5

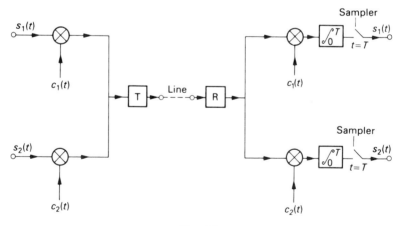

Fig. 4.6

* See F. R. Connor, *Noise*, Edward Arnold (1982).

used; this corresponds to integration over the bit period T, followed by appropriate sampling after time interval T.

4.4 Signal distortion

When the waveform of the received signal is not identical to the transmitted one, then it has been distorted during transmission. There are various types of distortion which a signal may suffer and these are generally classified as:

(a) Attenuation distortion

When the amplitude of the signal has been reduced, the signal has been attenuated. This is a common form of distortion and, if severe, it may lead to the complete disappearance of one or other of the frequencies contained in the original signal. Partly because of this, and partly because of noise, there is a limit to the distance over which a signal with a given power can be transmitted satisfactorily.

If the input power to a system or network is P_1 and the received power is P_2, then

$$\text{attenuation loss} = 10 \log_{10} \frac{P_1}{P_2} \, \text{dB}$$

If the overall attenuation of the system is being considered, then the attenuation is called the *transmission loss*.

(b) Phase distortion

In a complex signal, the various frequencies travel through the transmission medium with different velocities and so arrive at the receiving end in different phase or out of step. Such distortion is not serious in some communication systems, e.g. telephony, but can be in the case of television. The distortion is best understood by a consideration of the phase and group velocities of the signal transmitted.

Consider a signal of a single frequency f Hz being transmitted through a medium. Each crest of the wave travels forward with a *phase velocity* v_{ph}, such that successive peaks follow each other in a time T after travelling a distance λ. This is shown in Fig. 4.7.

$$v_{ph} = \frac{\lambda}{T} = \lambda f$$

If the phase shift per unit length is β, then over a wavelength of λ the phase shift is 2π radians.

Hence

$$\beta \lambda = 2\pi$$

or

$$\lambda = \frac{2\pi}{\beta}$$

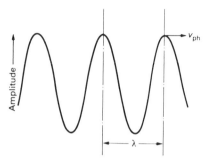

Fig. 4.7

Now
$$v_{ph} = \lambda f = \frac{2\pi f}{\beta}$$

or
$$v_{ph} = \frac{\omega}{\beta}$$

Suppose now the signal contains two frequencies ω_1 and ω_2 close to one another and travelling in the x-direction in a *dispersive* medium, i.e. a medium in which waves of different frequencies travel with different phase velocities. The phase-change coefficients β_1 and β_2 respectively are slightly different and the resultant pattern is similar to that produced by 'beats', in which the envelope passes successively through maxima and minima.

Let the equations of motion for the two waves travelling in the x-direction be given by the voltages V_1 and V_2 where

$$V_1 = A \sin(\omega_1 t - \beta_1 x)$$
$$V_2 = A \sin(\omega_2 t - \beta_2 x)$$

where $(\omega_1 - \omega_2) = \delta\omega$ (if $\omega_1 > \omega_2$) and $(\beta_1 - \beta_2) = \delta\beta$ (if $\beta_1 > \beta_2$).
The resultant voltage V is given by

$$V = V_1 + V_2 = A\left[\sin(\omega_1 t - \beta_1 x) + \sin(\omega_2 t - \beta_2 x)\right]$$

or
$$V = 2A \cos\left\{\frac{\delta\omega}{2}t - \frac{\delta\beta}{2}x\right\} \sin\left(\left\{\frac{\omega_1 + \omega_2}{2}\right\}t - \left\{\frac{\beta_1 + \beta_2}{2}\right\}x\right)$$

The resultant is a sine wave of frequency $[(\omega_1 + \omega_2)/2]$ and phase-change coefficient $[(\beta_1 + \beta_2)/2]$. Its amplitude varies cosinusoidally due to the factor $\cos\{(\delta\omega/2)t - (\delta\beta/2)x\}$ and is shown in Fig. 4.8 as the dotted envelope.

If the peak of the envelope at P is defined as the point x at time t, then V must be at its maximum value.

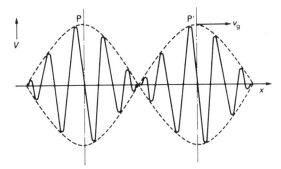

Fig. 4.8

Hence
$$\cos\left(\frac{\delta\omega}{2}t - \frac{\delta\beta}{2}x\right) = 1$$

or
$$\frac{\delta\omega}{2}t - \frac{\delta\beta}{2}x = 0$$

and
$$\frac{x}{t} = \frac{\delta\omega}{\delta\beta} = \frac{d\omega}{d\beta} \quad \text{(in the limit)}$$

This expression represents a velocity and so the peak of the envelope travels forward with a *group velocity* v_g given by

$$v_g = \frac{d\omega}{d\beta}$$

The waves move forward as a 'group' with this velocity. If the group velocity is not constant, the waves get out of step and phase distortion occurs. For no phase distortion to occur, $d\omega/d\beta$ must be constant and this means that the phase shift must vary linearly with frequency, i.e. there must be a linear phase-shift characteristic.

(c) Harmonic distortion

This is the distortion usually associated with amplifiers. Due to the non-linear characteristics of valves and transistors, new frequency components are produced which were not present in the original signal. Examples of this are second- and third-harmonic distortion and it is usual to express them as percentages of the fundamental. A small percentage of harmonic distortion is unavoidable and can be tolerated.

(d) Intermodulation distortion

When a signal containing two or more frequencies is passed through a non-linear device, e.g. a power amplifier, new sum and difference frequencies are produced due to 'modulation' between the frequency components in the original signal. This intermodulation distortion can be objectionable as the new frequencies may not be harmonically related to the original frequencies and, in the case of music, will sound displeasing to the ear. Third-order and fifth-order intermodulation products are usually the most troublesome in practice.[17]

(e) Crosstalk

In multichannel systems, interference between one channel and another is called crosstalk. Such crosstalk arises in an FDM system if the sidebands fall within an adjacent channel due to insufficient channel bandwidth, or it may be caused by intermodulation products produced by carriers passing through amplifiers in the transmission path.

In TDM systems, crosstalk is due to the overlap between adjacent channel pulses which arises due to the poor characteristics of the transmission path. Intelligible crosstalk as in telephone circuits is most objectionable, while unintelligible crosstalk in broadband carrier systems behaves like noise. Crosstalk effects are reduced to acceptable levels which, in practice, are down by about 60 dB.

(f) Digital errors

In the transmission of digital information, errors are bound to occur with a definite probability. Hence, in any message consisting of a certain number of digits, the error probability of a received message is of interest.

As an example, suppose a message is transmitted with a five-digit code. The probability of a digit being in error is p and the probability of it being correct is $(1-p)$. The error probability of the message for some cases will now be determined.

One-digit error message

The probability of one error is p and the probability of the remaining four digits being correct is $(1-p)^4$. Hence, the joint probability of one error digit and four correct digits is $p(1-p)^4$. This can occur in $^5C_1$ ways and so the total probability is $5p(1-p)^4 \simeq 5p(1-4p)$, since $p \ll 1$.

Two-digit error message

The probability of two error digits is p^2 and the probability of the remaining three digits being correct is $(1-p)^3$. Hence, the joint probability of two error digits and three correct digits is $p^2(1-p)^3$. This can occur in $^5C_2$ ways and so the total probability is $10p^2(1-p)^3 \simeq 10p^2(1-3p)$.

All-correct message

The probability of five digits being correct at the same time is $(1-p)^5$ and this can occur in only one way. The total probability is thus $(1-p)^5 \simeq (1-5p+10p^2)$.

Comments
1. Since p is usually about 10^{-4}, the probability of two or more error digits in a message is very small.
2. To overcome digital errors in transmission, and to maintain these errors at an acceptable level, which is usually called the bit error rate (BER), various error-detecting and error-correcting codes have been devised. Further details are given in Chapter 6.

Example 4.1
The anode current of a triode output stage with resistive load is given by

$$i_a = a + bv + cv^2$$

where v is the instantaneous voltage applied to the input.

Derive an expression for the ratio of second-harmonic to fundamental voltage in the output when the applied input voltage is sinusoidal. If a sine-wave input signal causes the anode current to vary between 1·7 and 0·45 of the quiescent value, calculate the percentage of second-harmonic in the output.   (U.L.)

Solution
Let the input voltage be given by

$$v = v_0 \sin \omega t$$

Hence

$$i_a = a + b(v_0 \sin \omega t) + c(v_0 \sin \omega t)^2$$

$$= a + bv_0 \sin \omega t + cv_0^2 \left(\frac{1 - \cos 2\omega t}{2} \right)$$

$$= a + bv_0 \sin \omega t + \frac{cv_0^2}{2} - \frac{cv_0^2}{2} \cos 2\omega t$$

and   percentage second-harmonic $= \dfrac{cv_0^2/2}{bv_0} \times 100 = \dfrac{cv_0}{2b} \times 100$

The quiescent current flows when $\omega t = 0$

or

$$I_0 = a + 0 + \frac{cv_0^2}{2} - \frac{cv_0^2}{2} = a$$

Maximum current flows when $\omega t = \pi/2$.

Hence $i_{max} = a + bv_0 + cv_0^2$

Minimum current flows when $\omega t = 3\pi/2$.

Hence $i_{min} = a - bv_0 + cv_0^2$

Combining i_{max} and i_{min} yields

$$i_{max} + i_{min} = 2a + 2cv_0^2 = 2I_0 + 2cv_0^2$$

and $i_{max} - i_{min} = 2bv_0$

with
$$\frac{(i_{max} + i_{min} - 2I_0)}{(i_{max} - i_{min})} = \frac{2cv_o^2}{2bv_o} = \frac{cv_o}{b}$$

Dividing through by I_0 gives

$$\frac{\dfrac{i_{max}}{I_0} + \dfrac{i_{min}}{I_0} - 2}{\dfrac{i_{max}}{I_0} - \dfrac{i_{min}}{I_0}} = \frac{cv_o}{b}$$

Substituting the numerical values leads to

$$\frac{(1\cdot7 + 0\cdot45) - 2}{(1\cdot7 - 0\cdot45)} = \frac{cv_o}{b}$$

or

$$\frac{0\cdot15}{1\cdot25} = \frac{cv_o}{b}$$

Hence percentage second-harmonic distortion $= \dfrac{cv_o}{2b} \times 100 = \dfrac{0\cdot15}{2 \times 1\cdot25} \times 100$

$$= \frac{15}{2\cdot5} = 6\,\%$$

5

Signal processing

5.1 Sampling of signals

The usual way of transmitting an analogue signal is for the entire signal to be transmitted continuously. However, on closer examination, it has been found that, provided certain conditions are met, it is sufficient to transmit only 'samples' of the analogue signal at given intervals of time. The original signal can still be recovered at the receiving end. The technique is called 'sampling'.[18]

In order to understand this novel idea, consider a train of pulses with a repetition frequency f_r and period T where $f_r = 1/T$. Assume that the analogue signal and its spectrum are of the forms shown in Fig. 5.1 where W is the highest frequency component in the spectrum.

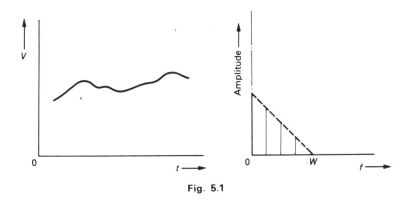

Fig. 5.1

If the analogue signal is used to amplitude-modulate the pulse train, the resultant waveform will consist of pulses whose amplitudes are, in effect, samples of the analogue signal at definite time intervals of T, as shown in Fig. 5.2. In *impulse* sampling, the sampling pulses are considered as extremely narrow (delta pulses) and the sample values follow the signal waveform at the sampled points. In *finite-width* sampling, which is more usual in practice, the samples are flat-topped at the various sampled points. However, in both cases the spectrum of the modulated pulse train is very similar, except for some high-frequency loss in the latter case which can be corrected by using equalisation.

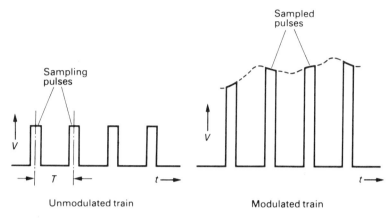

Fig. 5.2

An analysis of the amplitude-modulated pulse train reveals that its spectrum consists of the discrete Fourier components, $0, f_r, 2f_r, \ldots$, etc., of the unmodulated pulse train and that at each of these there is a set of sum and difference frequencies (called the upper and lower sidebands) due to each frequency component of the *analogue* signal. This is illustrated in Fig. 5.3.

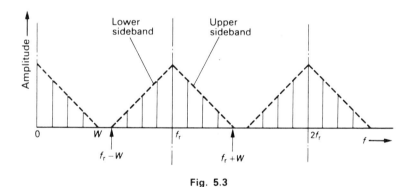

Fig. 5.3

At the zero-frequency component (d.c.), only sum frequencies are physically realisable and, provided W and $(f_r - W)$ do not overlap, it is possible to separate out the first group of frequencies at the receiving end. This is done by using a low-pass filter with a cut-off frequency $f_c = W$ and the separated frequencies are precisely those of the analogue signal transmitted. This leads to

the condition that

$$f_r - W \geqslant W$$

or

$$f_r \geqslant 2W$$

i.e. the repetition frequency must be at least equal to *twice* the highest frequency component in the analogue signal.

This condition imposes an upper limit on W, otherwise f_r would be excessively high and the pulses extremely narrow. It is generally difficult to generate very narrow pulses in practice and it also increases bandwidth.

The sampling technique can only be used in systems where the bandwidth of the sampled signal can be restricted to a maximum value W without destroying the essential information in the signal. This is done by first passing the analogue signal through a low-pass filter which removes all frequencies higher than W and this band-limited signal is used to modulate the pulse train.

Therefore, the second condition imposed by sampling is the use of *band-limited* signals and the sampling technique is expressed in a concise form in the following theorem.

5.2 Sampling theorem

Any function of time $f(t)$ whose highest frequency is W Hz can be completely determined by sampled amplitudes spaced at time intervals of $1/2W$ apart.

Proof
Suppose $F(\omega)$ is the complex spectrum of a *discontinuous* function $F(t)$, then

$$F(t) = \frac{1}{2\pi} \int_{-\infty}^{+\infty} F(\omega)\,e^{j\omega t}\,d\omega$$

The parameter t has finite values since $F(t)$ is a pulse, while ω extends to $\pm\infty$. Conversely, if ω is restricted to finite values $\pm 2\pi W$, t will extend to $\pm\infty$. Hence, let $F_T(\omega)$ be the *truncated* function of similar form to $F(\omega)$ over the interval $\pm 2\pi W$, as shown in Fig. 5.4, while $F(t)$ becomes the *periodic* function $f(t)$.

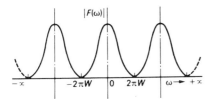

Fig. 5.4

Consider time t in intervals of $t = 1/2W$, then we have $nt = n/2W$ where $-\infty < n < +\infty$. Hence

$$f(t) = f(n/2W) = \frac{1}{2\pi} \int_{-2\pi W}^{+2\pi W} F_T(\omega)\, e^{j\omega(n/2W)}\, d\omega$$

$$= \frac{1}{2\pi} \int_{-2\pi W}^{+2\pi W} F(\omega)\, e^{j\omega(n/2W)}\, d\omega$$

since $F_T(\omega) \equiv F(\omega)$ over the interval $-2\pi W$ to $+2\pi W$.

The expression on the right is similar *in form* to the Fourier coefficients C_{-n} of Section 2.4. It represents the nth coefficient of a series in ω, having a periodicity of $4\pi W$ and with $f(t)$ replaced by $F(\omega)$.

Hence
$$C_{-n} = \frac{1}{4\pi W} \int_{-2\pi W}^{+2\pi W} F(\omega)\, e^{j\omega(n/2W)} d\omega$$

This expression is identical with $f(n/2W)$ above except for a factor of $1/2W$.

Hence
$$C_{-n} = \frac{1}{2W}\, f(n/2W)$$

or
$$C_{+n} = \frac{1}{2W}\, f(-n/2W)$$

This result shows that the C_n coefficients which *completely* defined the periodic function $f(t)$ in Section 2.4 are here identical with the sampled values $f(\pm n/2W)$. It therefore follows that the sampled values are sufficient to define $f(t)$ completely.

5.3   Sampled response

Each sampled pulse can be regarded as sufficiently different from the following pulse and can be treated as a narrow rectangular pulse. The response of each pulse when passed through an ideal low-pass filter is a $(\sin x)/x$ curve as indicated in Section 3.2(c).

The sampled pulses will therefore produce a series of such responses each delayed by the sampling interval $1/2W$. The algebraic sum of the responses produces a resultant waveform similar to that of the analogue signal sampled.

For the purposes of illustration, assume that the analogue signal has a sinusoidal variation. The resultant is shown in Fig. 5.5 and a mathematical proof is given in Appendix D.

5.4   Communication codes[19]

Since the early days of telegraphy, signals have been sent in some form of code. A code is a particular arrangement of symbols or pulses which conveys a certain

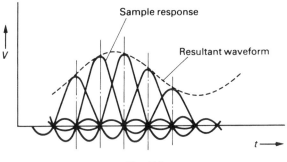

Fig. 5.5

character or message. More recently, the idea of signal sampling and the transmission of coded signals has been utilised in data communications with the aid of digital computers. Data is handled in digits (pulses) according to some pre-determined code, such as the binary code.

To achieve various objectives, many different types of codes have been designed, and brief descriptions of some well-known codes used in common practice will be given here. A more detailed study of coding theory and some important codes will be considered in the next chapter.

(a) Morse code (international)
Messages are sent by two elementary signals – a dot and a dash, as shown in Fig. 5.6.

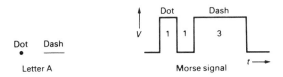

Fig. 5.6

The difference between the two signals is one of duration, being one unit of time for a dot and three units of time for a dash. The letters of the alphabet, numbers, and certain other characters are given combinations of the elementary signals. Spacing periods of time indicate the letters and words, i.e. space between letters is three units of time and between words it is five units of time.

However, the duration of time required for signalling each character is quite different, e.g. for the letter E it is a dot (1 unit) while for A it is a dot, followed by a dash (4 units). This makes it unsuitable for electromechanical operation but it is extensively used for hand-signalling as in amateur radio.

(b)   Five-unit code (teleprinter code)

This code is generally employed in teleprinter operation. It uses a basic five-unit code for all characters. Each character consists of a combination of five current pulses all of the same duration and known as space and mark. This is illustrated in Fig. 5.7.

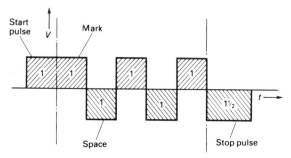

Fig. 5.7   Letter R

The code lends itself easily to machine operation because of the fixed time required for sending each character. However, to start the teleprinter and thus synchronise transmitter and receiver, a start pulse (1 unit) is sent before the character and a stop pulse (1·5 units) after the character, thus making it more correctly a 7·5-unit code.

Thirty-two combinations are initially possible and this is virtually doubled by using two of the combinations for transmitting 'case-shift' signals for numbers and letters. These are called 'figure-shift' and 'letter-shift' respectively, to indicate whether numbers or letters are being sent subsequently. The code is used extensively for automatic telegraphy.

(c)   Binary code

The code is based upon binary arithmetic and is used in both the fields of communications and computers. It is a two-digit code in which the digits are 'zero' and 'one'. As a signal, the 'zero' signifies no pulse while the 'one' signifies a pulse. These are usually called the binary digits, or simply 'bits' in communication language, and are illustrated in Fig. 5.8.

Various combinations of the digits can be used to denote the decimal numbers which are of prime importance in computation. Table 5.1 shows a typical arrangement for the numbers 0 to 10. Further details of binary arithmetic are given in Appendix E.

The basic technique when applied to communications is to convert the information or message into digital form. In the case of an analogue signal, e.g. a simple sine wave, the signal can be divided (quantised) into a set of levels, each designated by a decimal number, 0, 1, 2, etc., which is then expressed in binary digits as well. This is shown in Fig. 5.9.

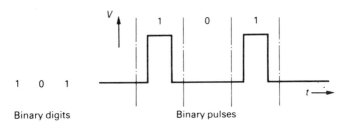

Fig. 5.8

Table 5.1

Decimal system	Binary system
0	0000
1	0001
2	0010
3	0011
4	0100
5	0101
6	0110
7	0111
8	1000
9	1001
10	1010

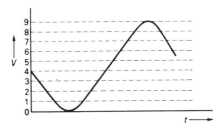

Fig. 5.9

Quantisation enables a complex signal to be transmitted in terms of the very elementary signals 0 and 1, in the form of a code. It forms the basis of pulse code modulation (PCM)* which is used in some communication systems. Figure 5.10 shows how it is used for a single-channel system.

* See F. R. Connor, *Modulation*, Edward Arnold (1982).

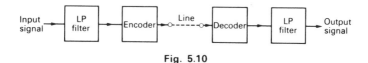

Fig. 5.10

The input signal is band-limited by a low-pass filter and then sampled. Each sampled pulse is converted into a group of pulses in the encoder, according to the binary code. The sets of pulses are transmitted to line, where they are decoded at the receiving end to produce the original sampled pulses. These sampled pulses, when passed through a low-pass filter, are converted back to the original message signal.

A variation of the binary code which also uses binary arithmetic is the *binary coded decimal* (BCD) code. Here, groups of binary digits are used to convey any particular decimal number, e.g. the number 9583 would be expressed as

$$1001\ 0101\ 1000\ 0011$$

and so it requires 16 digits in all.

Computers may employ the BCD code and conversion from binary to BCD or vice versa can be easily carried out by using the appropriate hardware.

(d)   Gray code[20]
For measuring rotation or position, mechanical analogue signals may be converted into digital form. The code changes from one number to the next by only one digit at a time and so avoids mechanical misalignment errors. A typical Gray code for the numbers 0 to 10 is given in Table 5.2.

Table 5.2

Decimal number	Gray code
0	0000
1	0001
2	0011
3	0010
4	0110
5	0111
6	0101
7	0100
8	1100
9	1101
10	1111

(e)   Barker codes[21]
In order to synchronise a stream of digital data in a data system, the data is usually preceded by a short interval of bits to provide a synchronising pulse for

accurate timing of the commencement of the following stream of data. A special class of binary codes, such as the Barker codes, is often used for this purpose.

The Barker codes are optimum in the sense that the peak autocorrelation function equals unity while the peak sidelobe level is $\leqslant 1/n$ where n is the length of the code. However, only a small number of these codes exist and no Barker codes of length greater than $n = 13$ have been found. Typical Barker code sequences are shown in Table 5.3.

Table 5.3

n	Code sequences	Sidelobe level (dB)
2	10,11	-6.0
3	110	-9.5
4	1101, 1110	-12.0
5	11101	-14.0
7	1110010	-16.9
11	11100010010	-20.8
13	1111100110101	-22.3

(f) ASCII code

The American Standard code (ASCII) is used as a standard for data communications. It employs a 7-bit code group with an eighth parity-check bit and is capable of defining 128 characters. Data transmission systems operate at speeds from about 300 bits/s to 9·6 kbits/s, over the public telephone network.

5.5 Speech processing[22, 23]

The speech signal from a speaker may be adapted electrically to the transmission channel of a listener. Four different techniques may be used for speech processing and these are amplitude compression, time compression, frequency compression, and analysis/synthesis techniques.

Speech is made up of a continuous signal which changes gradually over periods of 20–30 ms and the mechanism of speech production is described in Appendix F. Since speech is produced by slow, articulate movements, these correspond to a much lower information rate than that signified by the acoustics spectrum. Hence, it implies considerable redundancy in speech and voice coders or *vocoders* exploit this redundancy by analysing the waveforms to extract important information-bearing parameters.

Vocoders perform measurements on the spectral envelope of the speech signal and the parameters extracted are transmitted using a lower bandwidth than that normally required by the speech signal. At the receiver, the speech signal is synthesised with some degradation in quality, but no serious loss of intelligibility. Various types of vocoders can be used and the most recent *formant* vocoders use digitised speech.

Digital speech systems use either waveform coding techniques or analysis/synthesis techniques.

In waveform coding, the speech signal is sampled and the samples are coded to reproduce another set of samples comprising the same waveform at the receiving end. A typical example of this system employs PCM techniques and, at a transmission rate of about 20 kbits/s, some form of syllabic companded delta modulation is normally preferred. The quality obtained is inferior to that of the telephone system but is suitable for radio-telephony.

In the analysis/synthesis method, there is no attempt to reproduce the same waveform at the receiving end. Instead, using a theoretical model, the signal is analysed to derive the parameters of this model. The parameters are transmitted to the receiving end in coded, digital form where they are used to control a speech synthesiser corresponding to the model used in the analysis.

In a certain discrete-time model, the responses of the vocal tract are represented by a feedback transversal filter. The three most significant resonances of the vocal tract can be modelled by a six-coefficient feedback transversal filter. Alternatively, the vocal tract resonances can be removed from the speech signal by means of a feed-forward transversal filter. The model assumes that the larynx pulse is a single sample pulse, whereas it is a triangular waveform. Moreover, the vocal tract resonances do not have the same shape of frequency response as those of the model.

In addition to the redundancy which the speech signal contains in the form of vocal tract resonances, there is, during voiced sounds, the redundancy due to the similarity between the waveform resulting from one larynx pulse and that resulting from the next. Various systems have been devised to remove both types of redundancy and these involve *linear predictive coding*.[24, 25]

The linear predictability of speech signals is based on a linear model of speech production. In the model shown in Fig. 5.11, the excitation of speech sounds is represented by a pulse generator for voiced sounds and by a noise generator for

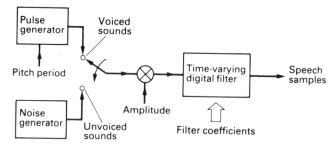

Fig. 5.11

unvoiced sounds. The vocal tract and radiation effects are represented by a linear, time-varying, digital filter.

The model provides a simple and effective way of representing speech signals in terms of a small number of slowly-varying parameters, such as pitch period, voiced and unvoiced excitation, amplitude control, and the various coefficients of an all-pole digital filter. The parameters are chosen such that the resulting output has the desired speech-like properties. For example, in a typical development, synthetic speech has been produced to provide a talking learning aid for young children.

5.6 Image processing[26]

Optical images may be processed electrically for various reasons, amongst which are data compression to reduce the time–bandwidth product, reduction of noise effects during transmission, improvement of a noisy or distorted image, and the extraction of particular features from an image. The link between the optical and electrical processes is given in Appendix F.

Data compression techniques generally exploit the perceptual redundancy in images which is a feature of the human visual process or the statistical redundancy produced by correlation between the picture elements in the image. Typical methods which have been used to implement data compression are transform coding, run length coding, line-to-line correlation, and directory look-up.

Processing to reduce the effects of noise may take place before or after transmission. A data compression process can be combined with an error-correcting code to improve the noise performance. On a typical facsimile image, a compression system giving a compression ratio of about 3 to 1 can be achieved by using a code which requires a $10–15\%$ increase in the data rate.

In many cases, the desired aim may not be to reconstruct the original image, but rather to render visible, details which have been obscured by noise or distortion. The optimum strategy may be to design the enhancement process to select the required features of the image, such as contrast enhancement, spatial frequency filtering, equalisation of the density probability distribution, or application of non-linear transfer functions.

Television processing[27]

In the PAL colour television system, a bit rate of 100–200 Mbits/s is required with error correction. Two possible methods of coding which may be employed to reduce costs are differential pulse code modulation (DPCM) and the Walsh–Hadamard transformation.

DPCM is a technique which exploits implied redundancies by use of a transmission code which represents the difference between each sample of a video signal. For a monochrome signal, previous-sample prediction may be used and this is shown in Fig. 5.12(a). A two-dimensional predictor with one

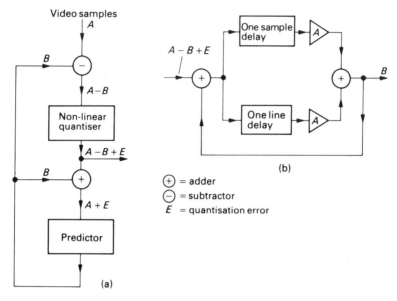

Fig. 5.12

sample delay and one line delay improves the prediction for vertical and horizontal transitions and is shown in Fig. 5.12(b).

For composite colour signals, previous-sample prediction is likely to cause overload when there are large and fast changes in the video signal due to chroma. As the colour information is conveyed on a high-frequency subcarrier at 4·43 MHz, one method of accurate coding of the colour subcarrier components is to employ a sampling frequency of three times the colour subcarrier frequency. By using third-previous-sample prediction, coding of a PAL 625-line system requires 1·5 less bits per sample than PCM coding. Here, again, prediction is improved by combining this with previous-line prediction.

In the Walsh–Hadamard technique, groups of n samples of the data are transformed to give the amplitudes of n Walsh functions. The Walsh function amplitudes corresponding to a set of sample values are conveniently obtained by matrix algebra using a particular type of Hadamard matrix. A Hadamard matrix is an orthogonal matrix whose normalised elements can only take on the values of ± 1. Walsh–Hadamard matrices may be defined by the recurrence equation

$$H_{2n} = \begin{vmatrix} H_n & H_n \\ H_n & H_{-n} \end{vmatrix}$$

where $H_n = 1$ and $H_{-n} = -1$ when $n = 1$.

In a real-time Walsh–Hadamard transformer, the input signal is obtained from an eight-bit PCM coder with a sampling frequency of about 13 MHz. Blocks of 32 sample values are transformed to give the amplitudes of the 32 corresponding Walsh functions which are fed to a bit rate reduction unit and then inverse-transformed to recover the sample values. The instrumentation of the transformer is based on a fast transform algorithm which is shown in Fig. 5.13 for blocks of 8 data samples.

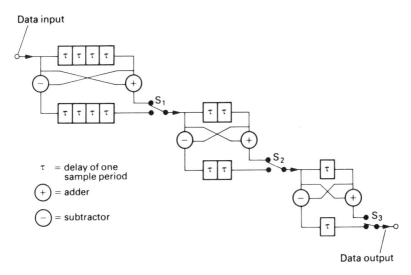

Fig. 5.13

The number of bits needed to represent the transform can be reduced considerably by removing some of the most significant bits from the coefficients. For the PAL 625-line spectrum, Walsh–Hadamard transformation requires 1 bit *more* per sample than DPCM but it is less immune to the effects of transmission errors than DPCM if no error-correction technique is employed.

Picturephone processing[28]
For the particular application of coding monochrome, picturephone signals consisting of 319 lines, 25 frames/s with 2:1 interlacing, both intraframe and interframe redundancy reduction techniques may be employed. Using the principle of DPCM, the source encoder quantises a differentiated version of the signal using a companded quantising law. This is an example of *intraframe* coding since it operates entirely within each frame and exploits spatial redundancy only.

A high-quality picturephone signal can be transmitted by DPCM at a data rate of 8 Mbits/s although, by exploiting temporal aspects of visual perception, the rate can be typically reduced by a factor of four. The signal still possesses statistical redundancy and sophistication might allow a nearer approach to the true information rate (based on a fidelity criterion) by using variable-length codes.

The greatest potential for data compression lies in *interframe* source coding. A substantial compression factor can be gained by transmitting data to update the store only for those picture elements which are deemed to be in moving areas. However, there are serious limitations imposed by interframe coding of picturephone signals which have yet to be resolved. These limitations arise in the efficiency with which noise-generated differences may be rejected without loss of genuine low-contrast movement. A data rate of 2 Mbits/s for picture-phone coding by interframe techniques has been demonstrated.

To operate the channel with a finite error rate, it is necessary to add redundancy in a suitable form as a channel code, and one of a wide range of techniques, e.g. a simple parity check or a convolutional code, may be used for this. Assuming a random-noise model for the channel, a simple 10-bit parity word per picturephone scanning allows correction of single-digit errors for 8 Mbits/s transmission with an error rate of 10^{-7}.

6

Information theory

The development of information theory[29-31] is due largely to the work of Hartley and Shannon which forms the basis of a more general field of theory referred to as communication theory. Though it is intended here to discuss only the basis of information theory, no discussion can begin without some definition of the words 'communication' and 'information'.

From a human point of view, the word 'communication' conveys the idea of one person talking or writing to another in words or messages. Language is the medium by which this is done, through the use of words derived from an alphabet. Not all words are used all the time and this implies that there is a minimum number which could enable communication to be possible. In order to communicate, it is necessary to transfer information to another person or, more objectively, between persons or machines.

This naturally leads to the definition of the word 'information', and from a communication point of view it does not have its usual everyday meaning. Information is not what is actually in a message but what *could* constitute a message. The use of the word *could* implies a statistical definition in that it involves some selection of the various possible messages. The important quantity is not the *actual* information content of a message but rather its *possible* information content.

This is the quantitative definition of information and so it is measured in terms of the number of possible selections that could be made. Hartley[29] was the first to suggest a logarithmic unit as a measure of information and this is given in terms of a message probability. The choice of a logarithmic unit as a measure of information has both practical and mathematical significance; it is found to retain the advantages of a linear relationship and at the same time to have concise mathematical form.

6.1 Average information *H*

The probability of transmitting a message before a message is received is called the *a priori* probability. Similarly, the probability of having transmitted a message after the message has been received is called the *a posteriori* probability.

Since the greater the *a posteriori* probability, the greater is the information received, we may define the information received from a message in terms of a logarithmic unit based upon the ratio of these two probabilities, i.e.

$$\text{information } I \text{ received} = \log\left[\frac{a \text{ } posteriori \text{ probability}}{a \text{ } priori \text{ probability}}\right]$$

Consider now a discrete message source producing a finite number of message symbols (pulses) with different probabilities, P_i, P_j, ..., P_m. If no noise is present, the transmitted message symbols can be received with certainty. Then the *a posteriori* probability is 1 and the *a priori* probability for the ith symbol is P_i, or

$$I = \log\frac{1}{P_i} = -\log P_i$$

which is called the *self-information*.

In the case of a very *large* number of n symbols, the ith symbol will occur nP_i times and so the information received from the source is the summation of all such quantities over the various values of i, or

$$\text{total } I = \sum_{i=1}^{m} -nP_i \log P_i = -n\sum_{i=1}^{m} P_i \log P_i$$

and the average information is given by

$$H = \frac{-n\sum_{i=1}^{m} P_i \log P_i}{n} = -\sum_{i=1}^{m} P_i \log P_i \text{ per symbol}$$

If there is a probability of receiving the same symbol every time, i.e. $P_i = 1$, a certainty, then

$$H = -\sum_{i=1}^{m} 1 \times \log 1 = 0$$

and so the measure of information is related to its *uncertainty*; the greater the uncertainty, the greater is the information conveyed.

Shannon was the first to show that information can be reduced to a binary system, i.e. a yes or no, representing a choice between two equal alternatives. These can be represented by the binary numbers 0 or 1 and are called binary digits or 'bits'. The unit of average information H is the bit and the logarithmic base is 2.

$$H = -\sum_{i=1}^{m} P_i \log_2 P_i \text{ bits/symbol}$$

If the number of symbols transmitted per second is n', then the number of bits of information transmitted per second is H' where

$$H' = n'H \text{ bits/s}$$

As an example, suppose we wish to transmit four different messages which are equally likely to occur. By proper coding we do not require four bits of information but only two, e.g. using the two binary digits 0 and 1, the messages are

$$00, \quad 01, \quad 10, \quad 11$$

i.e.

$$2^2 = 4$$

or

$$H = \log_2 4 = 2 \text{ bits}$$

In general, with b bits, we can obtain 2^b choices equal to some number N, such that

$$2^b = N$$

or

$$b = \log_2 N$$

or

$$H = \log_2 N \text{ bits}$$

The results obtained above for a discrete source also apply to a continuous source, since information from a continuous source could be received by sampling at discrete time intervals (sampling theorem) if the message signal is band-limited.

Example 6.1

Determine the average and maximum information H of a discrete, noiseless, binary channel if the probabilities of transmitting the two symbols 0 and 1 are p and $(1 - p)$ respectively.

Solution

The average information H is given by

$$H = \sum_{i=1}^{i=2} p_i \log_2 p_i \text{ bits/symbol}$$

where

$$H = -[p \log_2 p + (1 - p) \log_2 (1 - p)]$$

Converting from base 2 to base e, we obtain

$$H = -\log_2 e[p \ln p + (1 - p)\ln (1 - p)]$$

The maximum value of H is obtained by putting $dH/dp = 0$ which yields

$$\frac{dH}{dp} = -\log_2 e[1 + \ln p - 1 - \ln(1 - p)] = 0$$

or

$$\ln p = \ln(1 - p)$$

with

$$p = (1 - p)$$

or

$$p = \tfrac{1}{2}$$

Hence

$$H_{\text{max}} = -\log_2 e \left(\tfrac{1}{2}\ln\tfrac{1}{2} + \tfrac{1}{2}\ln\tfrac{1}{2} \right)$$

with $\qquad H_{max} = \log_2 e \times \ln 2$

or $\qquad H_{max} = 1$ bit/symbol

as illustrated in Fig. 6.1.

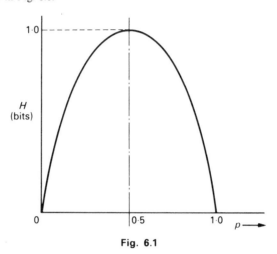

Fig. 6.1

6.2   Channel capacity C

The expression H for the average information output of a source has the same form as that for entropy in statistical mechanics. Hence, it is often referred to as the entropy of the source. In particular, if the probabilities P_i represent symbol probabilities, then H is the average information per symbol, as shown in Section 6.1.

In general terms, entropy refers to the randomness or disorder of a system; the greater the randomness, the greater is the entropy. For a communication system involving probabilities, entropy signifies choice or uncertainty. Hence, the greater the entropy of the system, the greater is the choice or uncertainty involved. When the entropy is zero, this corresponds to a certainty.

In the case of a noiseless channel, the transmitted and received signals are the same and have the same entropy. In a noisy channel, they are different and the entropy gives a measure of the uncertainty present due to noise.

Using the definitions of Shannon, let X and Y refer to the transmitter output and receiver input respectively. Then $H'(X)$ represents the uncertainty or entropy rate of the transmitted signal, $H'(Y)$ represents the uncertainty or entropy rate of the received signal, $H'(X|Y)$ represents the uncertainty or conditional entropy of the transmitted signal when the received signal is known, and $H'(Y|X)$ represents the uncertainty or conditional entropy of the received signal when the transmitted signal is known. Shannon then showed that the rate of transmitting information R is given by*

$$R = [H'(X) - H'(X|Y)] = [H'(Y) - H'(Y|X)]$$

where $H'(X)$, $H'(Y)$, $H'(X|Y)$, and $H'(Y|X)$ are in bits/s.

The channel capacity C is defined as the amount of information *correctly* transmitted per second and is given in bits/s. In any transmission system, noise is always present and must be taken into account. As noise leads to an uncertainty in the information received, it reduces the error-free capacity of a noisy channel. This means that R is equal to the rate of transmitting information $H'(X)$ less the uncertainty of what was sent $H'(X|Y)$ or, alternatively, it is the information received $H'(Y)$ less the uncertainty due to noise $H'(Y|X)$. The uncertainty $H'(X|Y)$ Shannon calls the 'equivocation'. It follows that the maximum channel capacity C is the maximum value of R. Hence

$$C = \text{maximum of } [H'(X) - H'(X|Y)]$$

This result follows from Shannon's theorem which states that, if the entropy rate R is equal to or less than the capacity C, there exists a coding technique which enables transmission over the channel with an arbitrarily small frequency of error or $R \leqslant C$ and this restriction holds even in the presence of noise in the channel.

A converse to this theorem states that it is not possible to transmit messages without errors if $R > C$. Thus, the channel capacity is defined as the maximum rate of reliable information transmission through the channel.

For the communications engineer, the value of C for a continuous signal in a noisy channel is useful in many cases. The various levels of the signal can be represented by combinations of the binary digits 0 and 1.

If S is the signal power and N is the noise power, the total voltage level is $\sqrt{S+N}$ and the minimum level $\sqrt{N}$. Because of noise, there is an uncertainty and no two levels can be distinguished if they are closer than $\sqrt{N}$. Hence, we can only use $\sqrt{(S+N)/N}$ distinguishable levels. The information content H is then given by

$$H = \log_2 \sqrt{\frac{S+N}{N}} \text{ bits}$$

If the time taken to send a pulse is t, then from telegraphy it is known that the corresponding bandwidth required is W where

$$t = 1/2W$$

Hence, the maximum number of pulses, i.e. symbols that can be sent per second, is n where

$$n = 1/t = 2W \text{ pulses or symbols}$$

---

* See Appendix H.

or $\qquad$ maximum of $H' = nH = 2W \log_2 \sqrt{\dfrac{S+N}{N}}$ bits/s

If the communication capacity is C, then

$$C = \frac{H}{T} = \text{maximum of } H' = 2W \log_2 \sqrt{\frac{S+N}{N}} \text{ bits/s}$$

or $\qquad H = CT = WT \log_2 (1 + S/N)$ bits

This is the Hartley–Shannon law of information.

Note
Detailed expressions for the information rate and channel capacity of a continuous source are given in Appendix I.

Example 6.2
Obtain the communication capacity of a noiseless channel transmitting n discrete message symbols per second.

Solution
The communication capacity for a noisy channel is given by

$$C = \text{maximum of } [H'(X) - H'(X|Y)]$$

and for a noiseless channel the equivocation $H'(X|Y) = 0$. Hence, we obtain

$$C = \text{maximum of } [H'(X)]$$

In the case of n discrete message symbols, the average information (entropy) per symbol is given by

$$H(X) = - \sum_{i=1}^{i=n} P_i \log_2 P_i = \sum_{i=1}^{i=n} P_i \log_2 \frac{1}{P_i}$$

which was shown in Example 6.1 to be a maximum, for the case of the binary channel, when all the message probabilities P_i are equal, i.e. $P_i = 1/n$.

Hence, if n message symbols are transmitted per second, the information rate $H'(X) = nH(X)$ bits/s and the communication capacity is given by

$$C = \text{maximum of } [H'(X)] = \text{maximum of } nH(X)$$

with $\qquad C = n \left[n \times \dfrac{1}{n} \log_2 n \right]$

or $\qquad C = n \log_2 n$ bits/s

Example 6.3
Show, using the result of the sampling theorem, that the total information in binary units that can be transmitted in a communication channel in time T is given by

$$WT \log_2(1 + P_S/P_N)$$

where W is the channel bandwidth, P_S is the signal power, and P_N is the noise power in the channel.

In a certain 625-line television system, the picture is scanned 25 times per second and has an aspect ratio of $\frac{4}{3}$. If at any point in the picture the eye can perceive eight gradations of light intensity, determine the rate of transmission of information of the system. It may be assumed that the horizontal and vertical resolutions are equal and that the whole of the line scan is used for picture waveform. (U.L.)

Solution
By the sampling theorem, the repetition frequency f_r is given by

$$f_r \geq 2W$$

where W is the highest frequency in the signal sampled. Using the more practical limit of equality, we have

$$f_r = 2W$$

or $\qquad 1/f_r = 1/2W$ (this is the sampling period)

If t is the duration of a pulse, at least one pulse should be sent within a sampling period. Then

$$t \doteq \frac{1}{f_r} = \frac{1}{2W}$$

and the maximum number of pulses sent in T seconds is given by

$$T/t = 2WT \text{ pulses}$$

If the signal power is P_S and the noise power is P_N, the total voltage level is proportional to $\sqrt{P_S + P_N}$. Since levels closer than $\sqrt{N}$ cannot be distinguished because of the noise threshold

$$\text{number of usable levels} = \sqrt{(P_S + P_N)/P_N} = \sqrt{1 + P_S/P_N}$$

For a binary system, the total information sent in T seconds is $H'T$ where

$$H'T = nHT = nT \log_2 N \qquad \text{(from Section 6.1)}$$

and $nT = 2WT$ is the number of bits used (pulses) and N is the number of usable levels (choices).

Hence $\qquad H'T = 2WT \log_2 \sqrt{1 + P_S/P_N} = WT \log_2(1 + P_S/P_N) \text{ bits}$

Problem

$$\text{distance between lines} = w/625$$
$$\text{number of elements per line} = d/w \times 625 = \tfrac{4}{3} \times 625$$
$$\text{time for one line scan} = 1/(625 \times 25)$$
$$\text{time to scan picture element} = \frac{1/(625 \times 25)}{\tfrac{4}{3} \times 625} = 3/(625^2 \times 100)$$
$$\text{time to scan two picture elements} = 6/(625^2 \times 100)$$
$$\text{fundamental frequency} = \frac{625^2 \times 100}{6} = 6 \cdot 5 \text{ MHz} = W$$

and   number of usable levels = number of gradations = 8

or $\sqrt{1 + P_S/P_N} = 8$

Now   communication capacity = $H/T = 2W \log_2 \sqrt{1 + P_S/P_N}$

Hence $C = 2 \times 6.5 \times 10^6 \times \log_2 8$

or $C = 39 \times 10^6$ bits/s

6.3 Redundancy

If a message contains more symbols than are necessary to convey the information, such a message has redundancy. Any source which produces symbols that are dependent has redundancy since the choice of one symbol depends upon another.

The English language has considerable redundancy since the use of particular letters depends upon the choice of a previous letter. This means that, in a telegram, certain letters may be left out without destroying the true message. In other cases, as in the Hamming code, controlled redundancy is used to correct errors.

If the actual information (entropy) of a source is H and the maximum possible information (entropy) is H_{max}, the relative entropy or efficiency is H/H_{max} and the redundancy is defined as

$$\text{redundancy} = 1 - \frac{H}{H_{max}}$$

In the case of intersymbol dependence, the conditional entropy H_c is used instead of H and the redundancy is then given by

$$\text{redundancy} = 1 - \frac{H_c}{H_{max}}$$

Usually $H_c \leqslant H$ and the equality sign is true when the symbols are entirely independent. It can be shown that, for eight-letter groups, H_c for the English language is about 2 bits/symbol and, for the 26 letters of the alphabet plus a space, we obtain

$$H_{max} = \log_2 27 = 4.76 \text{ bits/symbol}$$

and $$\text{redundancy} = 1 - \frac{2}{4.76} \simeq 50\%$$

Example 6.4

A message source produces two independent symbols A and B with probabilities $P(A) = 0.3$ and $P(B) = 0.7$. Calculate the efficiency of the source and, hence, its redundancy.

If the symbols are received on average with two in every 100 symbols in error, calculate the transmission rate of the system.

Solution
Since the symbols are independent, the entropy of the source is given by

$$H = -\sum_i P_i \log_2 P_i = \sum_i P_i \log_2 \frac{1}{P_i}$$

$$= 0.3 \log_2 \frac{1}{0.3} + 0.7 \log_2 \frac{1}{0.7}$$

$$= 0.5224 + 0.362$$

$$= 0.8844 \ \text{bit/symbol}$$

Since the maximum entropy occurs when the symbol probabilities are equal

$$H_{max} = \log_2 2 = 1 \ \text{bit/symbol}$$

with $\qquad$ efficiency $= \dfrac{H}{H_{max}} = \dfrac{0.8844}{1} = 0.8844 = 88.44 \%$

and $\qquad$ redundancy $= 1 - \dfrac{H}{H_{max}} = 1 - 0.8844 = 11.56 \%$

Since two in every 100 symbols are in error, the conditional probabilities $P(A|B)$ and $P(B|A)$ of an A or a B being transmitted, when a B or an A was received in error respectively, are as shown below.

		$P(A\|B)$	$P(B\|A)$
Transmitted	A	0.98	0.02
	B	0.02	0.98

Hence $\qquad H(X|Y) = \sum_i \sum_j P(j) P(i|j) \log_2 P(i|j)$

or $\qquad H(X|Y) = -0.7[0.98 \log_2 (0.98) + 0.02 \log_2 (0.02)]$

$$-0.3[0.02 \log_2 (0.02) + 0.98 \log_2 (0.98)]$$

$$= \left[0.98 \log_2 \frac{1}{0.98} + 0.02 \log_2 \frac{1}{0.02} \right]$$

$$= 0.141 \ \text{bit/symbol}$$

As $H(X)$ is 0.8844 bit per symbol, and assuming that 100 symbols are transmitted per second, then

$$H'(X) = 0.8844 \times 100 = 88.44 \ \text{bits/s}$$

and $\qquad H'(X|Y) = 0.141 \times 100 = 14.1 \ \text{bits/s}$

with $\qquad R = H'(X) - H'(X|Y) = 88.44 - 14.1 \ \text{bits/s}$

or $\qquad\qquad R = 74.34 \ \text{bits/s}$

6.4   Coding theory[32,33]

The importance of coding information for communication purposes was mentioned earlier in connection with the channel capacity and there are two ways by which the characteristics of the source can be 'matched' to those of the channel. They are known as *source* coding and *channel* coding.

In source coding, the principle is to use the minimum number of bits required to convey the necessary information efficiently and subject only to the constraints of a fidelity criterion. Hence, the aim is to remove redundancy or intersymbol dependence as far as possible without destroying the content or nature of the information conveyed.

In channel coding, redundancy is added to combat the effects of transmission errors due to noise. It involves the use of extra check digits which are transmitted, in a coded form, with the information digits, and are known as error-correcting codes. Most of the development in coding theory has taken place in the area of channel coding; only recently has more work been done in the difficult area of source coding.

6.4.1   Source coding

Two familiar encoding techniques which illustrate this idea are the Fano code and the Huffman code, and these are described below. More complex techniques for speech and image signals were described in the previous chapter on signal processing.

Fano code

The messages are arranged in descending order of probability and grouped into nearly equiprobable groups. To each group is assigned the symbol 1 or 0. This is illustrated in the table on the next page for a source producing eight messages $m_1, m_2, \ldots, m_8$ with probabilities $P_1, P_2, \ldots, P_8$. Here

$$\text{entropy of source} = \sum_{i=1}^{m} P_i \log_2 1/P_i$$

$$= 2 \cdot 44 \ \text{bits}$$

Hence   efficiency $= 2 \cdot 44 / 2 \cdot 55 = 95 \cdot 5 \%$

Huffman code

The messages are arranged in a column in descending order of probability and the two lowest probabilities are grouped to form a new probability. The new probability and the remaining probabilities are placed in a new column in descending order as before. The procedure is repeated till the final probability is 1·0.

Each of the grouped probabilities forms a junction which is assigned the digit 1 or 0. The coding of any message is obtained by following its horizontal line and the appropriate arrow lines to the final point 1·0. The code digits at each

m	P	Coding sequence				Code	Bits
m_1	0·4	1			1	11	$2 \times 0·4$
m_2	0·2				0	10	$2 \times 0·2$
m_3	0·15		1		1	011	$3 \times 0·15$
m_4	0·10				0	010	$3 \times 0·10$
m_5	0·06			1	1	0011	$4 \times 0·06$
m_6	0·04	0			0	0010	$4 \times 0·04$
m_7	0·03		0		1	0001	$4 \times 0·03$
m_8	0·02			0	0	0000	$4 \times 0·02$
ΣP	1·00						Total = 2·55

junction along the path traced give the required code in left-to-right sequence. This is illustrated in the example on the next page from which we obtain

$$\text{entropy of source} = 2·44 \text{ bits}$$

Hence $\qquad\qquad$ efficiency $= 2·44/2·49 = 98\%$

6.4.2 Channel coding

Error-correcting codes may be used to detect and correct channel errors due to noise and are generally classified as *block* codes or *convolutional* codes. A block code is one in which the check digits are used to check the information bits in the block, while in a convolutional code they check the information bits in previous blocks. Typically, an (n, k) code contains a total of n bits, with k information bits and $(n - k)$ check bits.

A block code $[C] = [c_1 c_2 c_3 \ldots c_n]$ can be defined by a parity-check matrix H if $[H] \oplus [C]^\mathsf{T} = 0$ where the symbol $\oplus$ denotes modulo-2 arithmetic and $[C]^\mathsf{T}$ is the transpose of the *row* matrix $[C]$, i.e. $[C]^\mathsf{T}$ is a column matrix. If $[H] \oplus [C]^\mathsf{T} \neq 0$, it leads to an error in the codeword which may be detected by means of a *syndrome* $[S]$ which is a column matrix yielding a binary number which indicates the digit error position. For example, if $[C']$ is the received codeword,

let $[H] \oplus [C']^\mathsf{T} = [S]$ and, if $[S] \equiv \begin{bmatrix} 0 \\ 0 \\ 1 \end{bmatrix}$, there is an error in the first digit.

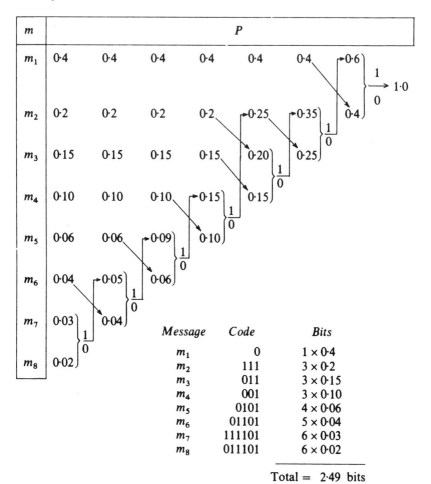

Message	Code	Bits
m_1	0	1×0.4
m_2	111	3×0.2
m_3	011	3×0.15
m_4	001	3×0.10
m_5	0101	4×0.06
m_6	01101	5×0.04
m_7	111101	6×0.03
m_8	011101	6×0.02

Total = 2·49 bits

Examples of such codes are the parity-check codes, the Hamming code, and the BCH code.

In the codewords used, it is convenient to refer to the Hamming weight $w(c_j)$ which is the number of 'ones' in the codeword c_j and the Hamming distance d_{jk} between two codewords which is the number of digit positions by which two codewords c_j and c_k *differ*. This is symbolically denoted by $d_{jk} = w(c_j \oplus c_k)$. For example, if the codewords are 00101 and 11100, the Hamming weights are 2 and 3 respectively and the Hamming distance is 3, i.e. $d_{jk} = w(00101 \oplus 11100) = 3$.

Parity-check code

The code can only detect a single error in a transmitted message. A parity check on the number of ones transmitted can be easily made at the receiver using odd or even parity. For example, a single-parity-check digit is added to the stream of message digits so that the number of ones transmitted is always even (even-parity-check). If a single error occurs in transmission, the number of received ones is odd, an error is at once detected, and a re-transmission of the message is requested.

Hamming code[34]

The code is used for detecting and correcting single errors in a series of received binary digits. A set of n digits is divided into a group of k information digits and c check digits or $n = k + c$. In order to check n positions, we must have the relationship $2^c - 1 \geqslant n$. Hence, redundancy or extra check digits are used to check errors, and for minimum redundancy $2^c - 1 = n$.

The parity-check matrix is easily constructed by using the binary equivalent of each column numbered as a decimal number (left to right). Hence, for four information digits and three check digits, $n = 7$ and $k = 4$. This yields a (7, 4) Hamming codeword with an H matrix of the form

$$[H] = \begin{bmatrix} 0 & 0 & 0 & 1 & 1 & 1 & 1 \\ 0 & 1 & 1 & 0 & 0 & 1 & 1 \\ 1 & 0 & 1 & 0 & 1 & 0 & 1 \end{bmatrix}$$

and a codeword $[C] = [c_1 c_2 k_1 c_3 k_2 k_3 k_4]$ where k_1, k_2, k_3, and k_4 are the information bits and c_1, c_2, and c_3 are the check bits satisfying the linear equations

$$c_1 = k_1 \oplus k_2 \oplus k_4$$
$$c_2 = k_1 \oplus k_3 \oplus k_4$$
$$c_3 = k_2 \oplus k_3 \oplus k_4$$

For example, suppose the codeword transmitted was 0010110 and it was received in error as 0010010 where the error is in the *fifth* position. To determine the syndrome $[S]$ indicating the error position in binary form we use modulo-2 arithmetic to obtain

$$[H] \oplus [C]^{\mathrm{T}} = \begin{bmatrix} 0 & 0 & 0 & 1 & 1 & 1 & 1 \\ 0 & 1 & 1 & 0 & 0 & 1 & 1 \\ 1 & 0 & 1 & 0 & 1 & 0 & 1 \end{bmatrix} \begin{bmatrix} 0 \\ 0 \\ 1 \\ 0 \\ 0 \\ 1 \\ 0 \end{bmatrix}$$

Hence
$$[H] \oplus [C]^T = \begin{bmatrix} 0 \oplus 0 \oplus 0 \oplus 0 \oplus 0 \oplus 1 \oplus 0 \\ 0 \oplus 0 \oplus 1 \oplus 0 \oplus 0 \oplus 1 \oplus 0 \\ 0 \oplus 0 \oplus 1 \oplus 0 \oplus 0 \oplus 0 \oplus 0 \end{bmatrix}$$

or
$$[S] = \begin{bmatrix} 1 \\ 0 \\ 1 \end{bmatrix}$$

and so the syndrome $[S] = [101]$ which is the binary code for the decimal number 5, confirming that the error is in the fifth position.

Bose–Chaudhuri–Hocquenhem code (BCH code)[35]

An example of a block code is the (n, k) BCH code in which the first set of k bits are the information bits and the last set of $(n - k)$ bits are the check bits. It is also a cyclic code and so codewords can be easily generated using a shift register as an encoder.

To decode a BCH code, a syndrome is formed by passing the received stream of bits into a shift register similar to the one at the transmitter. A bit by bit comparison of the check bits reveals if an error has occurred and yields a syndrome which is a record of the error pattern received. The decoder makes corresponding corrections to the information bits and the corrected information bits are read out by the decoder. An encoder and decoder for an (n, k) BCH code is shown in Fig. 6.2.

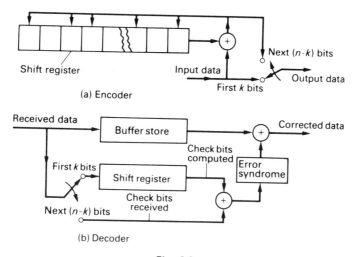

(a) Encoder

(b) Decoder

Fig. 6.2

BCH codes are widely used for random-error correction as they require the smallest number of check bits for a given reliability. The codes are constructed so that $n = 2^m - 1$ where m is an integer. They contain $m \times e$ parity-check bits which can correct up to e errors and detect up to $2e$ errors. For example, a (127, 106) BCH code with $n = 127$, $m = 7$, and $e = 3$ can correct up to 3 errors and detect up to 6 errors in any given block.

Convolutional codes[36]

In these codes, a continuous stream of bits is transmitted instead of being coded in blocks. The check digits are used to check overlapping groups of bits known as the constraint span. Such codes are generated by an n-stage shift register connected in a pre-determined way using feedback.

A convolutional code can be generated by combining the outputs of the shift register with one or more modulo-2 adders. The shift register is operated by clock pulses synchronised to the bit stream and, if n is the number of output bits per input bit, the code rate is $1/n$.

Decoding of the received bit stream does not require any block synchronisation and is achieved by checking the received sequence bit by bit and making a decision as to whether a 1 or 0 was transmitted. A decoding algorithm which may be employed is known as Viterbi decoding.[37] An example of an encoder using an n-stage shift register and a single modulo-2 adder is shown in Fig. 6.3.

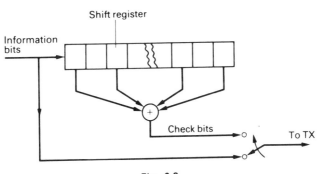

Fig. 6.3

Problems

1 A simple facsimile transmitter consists of a cylindrical drum mounted on a threaded rod. A light-emitting source illuminating a small area of the drum surface therefore scans along a helical path as the drum is rotated. The reflected light is sensed by a photodiode which detects changes in the pattern of a picture wrapped round the drum. The drum has a diameter of 12 cm, the picture to be transmitted has dimensions of 45 cm × 37·7 cm, and the drum rotates at 100 rev/min. Calculate the highest significant component that need be transmitted for a scanning density of 50 lines/cm. Also calculate the time taken to transmit the picture. State any assumptions made. (C.E.I.)

2 The following data applies to a television broadcast system.

number of scanning lines per picture	625
number of fields per second	50
interlace ratio	2/1
aspect ratio	4/3
$\dfrac{\text{duration of picture signal}}{\text{line scan time}}$	0·8

Calculate the frequency of
(a) the line time base,
(b) the field time base,
(c) the fundamental video frequency,
(d) the bandwidth of the system.

3 A radar transmitter is modulated by pulses 1 μs in duration at a frequency of 500 Hz. What is
(a) the pulse repetition frequency,
(b) the interpulse time,
(c) the duty cycle,
(d) the average power transmitted if the peak power is 1 kW?

4 A train of rectangular pulses of 10 V peak value have each a duration of 4 μs and there are 1000 pulses per second. Obtain the general term in the Fourier series and find the peak value of the 5 kHz component. Sketch the spectrum of the pulses and state the lowest harmonic of 1 kHz at which there will be no component. Indicate how the spectrum will be modified if the pulses are modulated by a 250 Hz sinusoidal signal. (U.L.)

Problems   101

5   The Fourier series of a function $f(t)$ of period 2π does not contain sine terms. Also $f(t) = \cos t$ when $0 \leqslant t \leqslant \pi/2$ and $f(t) = 0$ when $\pi/2 \leqslant t \leqslant \pi$. Sketch the waveform of $f(t)$ for one complete period and show that the Fourier series is

$$\frac{1}{\pi} + \frac{1}{2}\cos t + \frac{2}{\pi}\sum_1^\infty \frac{(-1)^{n+1}}{4n^2 - 1}\cos 2nt$$

6   One form of the Fourier integral which transforms the time function $f(t)$ to the frequency spectrum $F(\omega)$ is

$$F(\omega) = \frac{1}{\sqrt{2\pi}} \int_{-\infty}^{+\infty} f(t)\,e^{-j\omega t}\,dt$$

Use this integral to derive an expression for the spectrum of a rectangular pulse of amplitude 1 V and duration 1 μs. Draw a graph showing the variation of the spectral amplitude with frequency and determine the lowest frequency at which the amplitude is zero.

The pulse is applied to the input of a low-pass filter with a cut-off frequency at 100 kHz. Describe briefly the main features of the spectrum and the waveform of the output pulse.   (U.L.)

7   Use the Fourier transform

$$F(\omega) = \int_{-\infty}^{+\infty} f(t)\,e^{-j\omega t}\,dt$$

to derive an expression for the spectral amplitude $F(\omega)$ corresponding to the pulse waveform

$$f(t) = 0 \qquad -\infty < t < 0$$
$$f(t) = E\,e^{-\alpha t} \qquad 0 < t < +\infty$$

Draw a graph showing the variation of the amplitude with frequency and determine
(a)   the value of $F(0)$,
(b)   the frequency at which the amplitude is 3 dB below the value of $F(0)$.

If the pulse is developed as a voltage across R show, by considering either the waveform function or the spectral function, that the energy dissipated is $E^2/2\alpha R$ joules.   (U.L.)

8   A circuit consists of a 300 V d.c. supply in series with a switch S_1, a capacitor of 1 μF, an inductance of 4 H, and a resistor of 3·2 kΩ. The resistor is shunted by a second switch S_2 and, initially, both S_1 and S_2 are open with the capacitor uncharged. If, at time $t = 0$, S_1 is closed with S_2 open, derive an expression for the current at time t seconds later.

If S_2 is subsequently closed at time $t = \pi/600$ second, find the rate of change of current immediately after S_2 is closed and the frequency of this current.   (U.L.)

9   Draw the circuit corresponding to the equations

$$L\,\mathrm{d}i_1/\mathrm{d}t = Ri_2 = V$$

$$Ri_2 + q/C = 0$$

$$i_1 + i_2 = \mathrm{d}q/\mathrm{d}t$$

If the voltage V is constant, find the differential equation satisfied by q. Show that q oscillates only if $L < 4CR^2$. If this condition is satisfied, and if $q = 0$ and $i_1 = 0$ at $t = 0$, show that

$$q = VC - (V/2\omega R)\mathrm{e}^{-t/2RC}(\sin \omega t + 2\omega RC \cos \omega t)$$

where $\omega^2 = (1/LC) - (1/4R^2C^2)$.

10 The Fourier transform of a pulse is given by

$$F(\omega) = 2\pi/\omega_0 \qquad |\omega| \leqslant 0{\cdot}5\,\omega_0$$

$$F(\omega) = 0 \qquad |\omega| > 0{\cdot}5\,\omega_0$$

Derive the corresponding time function and show that, if the pulse interval in a train of such pulses is $2\pi/\omega_0$, the pulses can be detected without mutual interference.

Discuss the factors that influence the choice of pulse shapes to be used for digital communication systems, including the cases of AM, FSK, and PSK.
(C.E.I.)

11 Determine the discrete Fourier transform (DFT) of a time function $f(t)$ which is given at eight sampling points by

$$f(t) = +1 \qquad (t = 0, 1, 2, 3)$$

$$f(t) = -1 \qquad (t = 4, 5, 6, 7)$$

12 Show how the convolution integral may be expressed in terms of the superposition of impulse responses.

A voltage source $v(t)$ is applied to a series RC circuit. If $v(t) = u(t) - u(t - t_0)$ where $u(t)$ is the unit step function and t_0 is a constant, determine the capacitor voltage by means of convolution, assuming that the capacitor voltage is initially zero.

13 A signal is of the form $f(t) = 4(\sin 800\,\pi t + \sin 1600\,\pi t)$ and it is sampled at a frequency of 2 kHz. Obtain the spectrum for the sampled signal and determine the Nyquist rate for $f(t)$.

If the sampled signal is passed through an ideal low-pass filter, give a suitable cut-off frequency of the filter for recovering $f(t)$.

14 Give a block diagram of a system for the generation of sampled data from an analogue baseband signal. If the highest baseband frequency is 15 kHz, suggest with reasons a suitable duration for each sample. Prove that the sample rate must be at least 30 kHz and explain the form of distortion known as *aliasing*.

Give the principles of two interpolation techniques for recovery of the analogue signal from the sampled information and sketch circuits to implement each method. (C.E.I.)

15 Define the *information content* of a message emanating from a source. Calculate the average information content per message for a source providing three messages with probabilities 0·6, 0·3, and 0·1.

Supposing that the messages were in fact the letters 'e', 'u', and 'q' and that 'q' was always followed by 'u', what would be the effect on the average information? How can the effect be described? (C.E.I.)

16 A black and white television picture contains 500 'dots' per line. Each 'dot' has an equal probability of being black or white and the eye can distinguish the dots at a maximum rate of change of eight times per second. Calculate the total information content of the signal if 10 000 lines are transmitted per second.

17 Comment briefly on the bandwidth requirements for the transmission of speech, facsimile, and colour television.

The Hartley–Shannon law

$$I = WT \log_2 (1 + P_S/P_N)$$

gives the relationship between the information I transmitted, the bandwidth W, the duration of transmission T, and the signal-to-noise power ratio P_S/P_N. It is proposed to send a picture 9 cm × 12 cm over a facsimile transmission system having a bandwidth of 3 kHz. The picture may be assumed to have 10 levels of intensity and to require a resolution of 50 lines per cm vertically and 50 quantised values per cm horizontally. Estimate the minimum transmission time

(a) with no noise in the system,
(b) with a signal-to-noise ratio of 15 dB. (C.E.I.)

18 In a binary symmetric channel, the probabilities of the input messages are $P(x_1) = 0·6$ and $P(x_2) = 0·4$. If the conditional probabilities $P(y_1 | x_1) = 0·8$ and $P(y_2 | x_1) = 0·2$, determine the mutual information and channel capacity.

19 A discrete source transmits six message symbols with probabilities of 0·3, 0·2, 0·2, 0·15, 0·1, and 0·05. Devise suitable Fano and Huffman codes for the messages and determine the average length and efficiency of each code.

20 A Hamming (7, 4) code has the following parity-check matrix

$$[H] = \begin{bmatrix} 0 & 0 & 0 & 1 & 1 & 1 & 1 \\ 0 & 1 & 1 & 0 & 0 & 1 & 1 \\ 1 & 0 & 1 & 0 & 1 & 0 & 1 \end{bmatrix}$$

Determine the correct codeword if the information bits are 1010. If a received codeword is 0100001, determine if an error has been made and, if so, what is the correct codeword?

Answers

1   1570 Hz,   22·5 s

2   15·625 kHz,   50 Hz,   8·14 MHz,   bandwidth $\simeq$ 8·0 MHz   (VSB transmission)

3   500 Hz,   1·999 ms,   1/2000,   0·5 W

4   0·08 V,   250th harmonic,   sidebands at $(1 \pm 0·25)$ kHz, $(2 \pm 0·25)$ kHz, etc.

6   1 MHz

    Output pulse amplitude is 0·2 V and the basewidth is 10 μs. See Fig. 3.7 for the waveform.

7   E/α, $\alpha/2\pi$

8   $0·25\,e^{-400t}\sin 300t$,   79·5 Hz

10   $f(t) = \dfrac{\sin(\omega_0 t/2)}{\omega_0 t/2}$

    Comments
    (a)   At time intervals of $(2\pi/\omega_0)$, the pulse 'tails' crossover at the zeros in antiphase and cancel out exactly. Thus, there is no mutual interference between time samples.
    (b)   To minimise the use of bandwidth in digital systems and also intersymbol interference (crosstalk), pulse shapes are not rectangular, but rounded at the edges. Since it is difficult to produce the ideal $(\sin x)/x$ shape, alternatives used are either cosine or cosine-squared in shape. Pulse types used include on–off, bipolar, or duobinary forms.

11   $F(0\Omega) = F(2\Omega) = F(4\Omega) = F(6\Omega) = 0$
    $F(1\Omega) = 0·25 - j0·8535$
    $F(3\Omega) = 0·25 - j0·1035$
    $F(5\Omega) = 0·25 + j0·1035$
    $F(7\Omega) = 0·25 + j0·8535$

12   $v_C(t) = 1 - e^{-t/RC}$ $(0 < t < t_0)$
    $v_C(t) = e^{-t/RC}\big[e^{-t_0/RC} - 1\big]$ $(0 < t_0 < t)$

13   1·6 kHz,   1 kHz

14   If a baseband signal is not restricted in bandwidth or is sampled at too low a sampling frequency, some of the higher frequency components overlap or 'fold-over' into the baseband signal causing a form of harmonic distortion known as *aliasing*.

Two interpolation techniques which may be used are
(a)   direct low-pass filtering,
(b)   sample and hold circuit, followed by low-pass filtering.

15   $H_{av} = 1 \cdot 2946$ bits
The average information is reduced and the effect is known as redundancy
R where $R = 1 - H_{av}/H_{max}$.

16   15×10^6 bits/s

17   45 s,   59·5 s

18   0·229 bits,   0·279 bits

19   *Fano code*
11,   10,   01,   001,   0001,   0000;   2·40 bits,   98·92 %
Huffman code
11,   10,   00,   001,   1101,   0101;   2·40 bits,   98·92 %

20   1011010,   0100101

References

1 MARTIN, J. *Telecommunications and the Computer.* Prentice-Hall (1976).
2 FREEMAN, R. L. *Telecommunication System Engineering.* John Wiley (1980).
3 KENNEDY, G. *Electronic Communication Systems,* Second Edition. McGraw-Hill (1977).
4 TURNER, L. W. (ed) *Electronic Engineer's Reference Book.* Newnes-Butterworth (1976).
5 SKOLNIK, M. I. *Introduction to Radar Systems,* Second Edition. McGraw-Hill (1980).
6 SCHWARTZ, M. *Information, Transmission, Modulation and Noise,* Third Edition. McGraw-Hill (1980).
7 *Reference Data for Radio Engineers,* Sixth Edition. Howard W. Sams & Co. (1975).
8 LENDER, A. *Institute of Electrical and Electronic Engineers, Spectrum,* **3,** 104–15, February 1966.
9 RABINER, L. R. and GOLD, B. *Theory and Applications of Digital Signal Processing.* Prentice-Hall (1975).
10 TRETTER, S. A. *Introduction to Discrete-Time Signal Processing.* John Wiley (1976).
11 RABINER, L. R. and RADER, C. M. *Digital Signal Processing.* IEEE Press (1972).
12 BLINCHIKOFF, H. J. and ZVEREV, A. I. *Filtering in the Frequency and Time Domains.* John Wiley (1976).
13 *Transmission Systems for Communications.* Bell System Laboratories (1964).
14 HALLIWELL, B. J. (ed) *Advanced Communication Systems.* Newnes-Butterworth (1974).
15 DIXON, R. C. *Spread Spectrum Systems.* John Wiley (1976).
16 GOLOMB, S. W. *Digital Communications with Space Applications.* Prentice-Hall (1964).
17 MANASSEWITSCH, V. *Frequency Synthesizers.* John Wiley (1976).
18 JERRI, A. J. *Proceedings Institute of Electrical and Electronic Engineers,* **65,** 1565–96, 1977.
19 FINK, D. G. (ed) *Electronics Engineers' Handbook.* McGraw-Hill (1975).
20 SHEINGOLD, D. H. and FERRERO, R. A. *Institute of Electrical and Electronic Engineers, Spectrum,* **9,** 47–56, September 1972.
21 JACKSON, W. (ed) *Communication Theory,* p. 273. Butterworth Scientific Publications (1953).
22 RABINER, L. R. and SCHAFER, R. W. *Digital Processing of Speech Signals.* Prentice-Hall (1978).
23 LIM, J. S. and OPPENHEIM, A. V. *Proceedings Institute of Electrical and Electronic Engineers,* **67,** 1586–604, December 1979.
24 *Selected Papers in Digital Signal Processing,* **2,** 79. IEEE Press (1976).

25 Case Studies in Advanced Signal Processing. *Institute of Electrical Engineers Conference Publication No. 180*, 1979.
26 WINTZ, P. A. Transform Picture Coding. *Proceedings Institute of Electrical and Electronic Engineers*, **60**, 809–20, 1972.
27 CAPELLINI, V. *et al. Digital Filters and their Applications*, Chapter 8. Academic Press (1978).
28 AAGAARD, E. A. *et al.* An Experimental Video Telephone System. *Philips Technical Review*, **36**, No. 4, 85, 1976.
29 HARTLEY, R. V. The Transmission of Information. *Bell System Technical Journal*, **3**, 535, 1928.
30 SHANNON, C. E. A Mathematical Theory of Communication. *Bell System Technical Journal*, **27**, 379–423 and 623–56, 1948.
31 SHANNON, C. E. Communication in the Presence of Noise. *Proceedings Institute of Radio Engineers*, **37**, 10–21, 1949.
32 PETERSON, W. W. and WELDON, E. J. *Error-Correcting Codes.* John Wiley (1972).
33 WIGGERT, D. *Error-Control Coding and Applications.* Artech House (1978).
34 HAMMING, R. W. Error-Detecting and Error-Correcting Codes. *Bell System Technical Journal*, **29**, 147–60, April 1950.
35 BOSE, R. C. and RAY-CHAUDHURI, D. K. On a Class of Error Correcting Binary Group Codes. *Information and Control*, **3**, 68–79 and 279–90, 1960.
36 FORNEY, G. D. *Institute of Electrical and Electronic Engineers, Spectrum*, **7**, 47–58, June, 1970.
37 VITERBI, A. J. Error Bounds for Convolutional Codes and an Asymptotically Optimum Decoding Algorithm. *Institute of Electrical and Electronic Engineers, Transactions on Information Theory*, **IT–13**, No. 2, 260–69, 1967.
38 WALSH, J. L. A Closed Set of Normal Orthogonal Functions. *American Journal of Mathematics*, **45**, 5–24, 1923.
39 SHANNON, C. E. Coding Theorems for a Discrete Source with a Fidelity Criterion. *Institute of Radio Engineers National Convention Record*, **4**, 142–63, 1959.
40 DAVISSON, L. D. Rate Distortion Theory and Application. *Proceedings Institute of Electrical and Electronic Engineers*, **60**, 800–9, July 1972.

Appendices

Appendix A: The Doppler effect

The whistle of a train appears to increase in pitch (frequency) as it moves towards a stationary observer and it appears to decrease in pitch as it moves away from the same observer. This is known as the Doppler effect and, in Doppler radar, a moving aircraft may transmit an electromagnetic wave and receive it back after reflection from the ground. Alternatively, a ground station radar may emit a similar wave and receive it back after reflection from a moving aircraft within its range. In both cases, a Doppler shift in frequency is observed which is used to determine the speed of the aircraft.

To evaluate this frequency shift, suppose a stationary aircraft at A emits an electromagnetic wave of frequency f, directly towards a stationary reflecting surface at S, for a time t and receives it back after reflection as shown in Fig. A.1(a). The number of waves emitted is ft and they cover a double distance 2AS and so $AS = ft\lambda/2$ where λ is the wavelength. If the aircraft also travels towards a point A' closer to S, with a velocity v, in the time t, then the ft waves emitted are squeezed into the shorter double distance 2A'S and so have a shorter wavelength λ' where $A'S = ft\lambda'/2$.

Hence

$$AS = A'S + vt$$

or

$$\frac{ft\lambda}{2} = \frac{ft\lambda'}{2} + vt$$

with

$$\lambda = \lambda' + \frac{2v}{f}$$

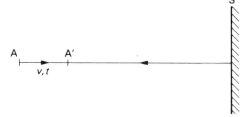

Fig. A.1(a)

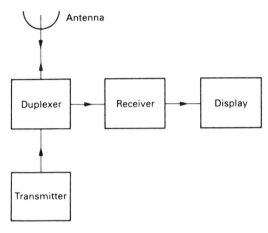

Fig. A.1(b)

or
$$\frac{\lambda}{\lambda'} = 1 + \frac{2v}{f\lambda'}$$

Since $f\lambda = f'\lambda' = c$, the velocity of the electromagnetic wave, and f' is the apparent frequency of the reflected wave, we obtain

$$\frac{f'}{f} = 1 + \frac{2v}{f\lambda'}$$

with
$$\frac{(f'-f)}{f} = \frac{2v}{f\lambda'}$$

If $(f'-f) = \Delta f$, the small Doppler frequency shift, then

$$\Delta f = \frac{2v}{\lambda'} = \frac{2vf'}{c} \simeq \frac{2vf}{c}$$

since $f \simeq f'$.

In Doppler radar, a CW signal is transmitted towards the ground and the received signal is mixed with part of the transmitted signal to obtain the Doppler shift Δf, which is then fed into a measuring device and displayed. For an aircraft flying horizontally above ground with its antenna pointing towards the ground at an angle of θ to the horizontal, the velocity of approach is $v\cos\theta$ and so

$$\Delta f = \frac{2vf\cos\theta}{c}$$

A typical arrangement is shown in Fig. A.1(b) for a single-beam system. In practice, several beams are used to provide more accurate velocity information.

Appendix B: Signum and impulse functions

Signum function

This function is defined by

$$f(t) = -1 \qquad -\infty < t < 0$$
$$f(t) = +1 \qquad 0 < t < +\infty$$

Hence, its Fourier transform is given by

$$F(\omega) = \int_{-\infty}^{+\infty} f(t) e^{-j\omega t} dt$$

$$= \int_{-\infty}^{0} -e^{-j\omega t} dt + \int_{0}^{+\infty} e^{-j\omega t} dt$$

$$= \left[\frac{e^{-j\omega t}}{-j\omega} \right]_{0}^{+\infty} + \left[\frac{e^{-j\omega t}}{-j\omega} \right]_{0}^{+\infty}$$

$$= \frac{1}{j\omega}[0+1] + \frac{1}{j\omega}[0+1]$$

or $$F(\omega) = \frac{2}{j\omega}$$

Impulse function (unit impulse)

This is the same as the Dirac delta function and is defined in the ω-plane by

$$\int_{-\infty}^{+\infty} \delta(\omega) = 1$$

Hence, its inverse Fourier transform is given by

$$f(t) = \frac{1}{2\pi} \int_{-\infty}^{+\infty} \delta(\omega) e^{j\omega t} d\omega$$

with $$f(t) = \frac{e^{j\omega t}}{2\pi} \int_{-\infty}^{+\infty} \delta(\omega) d\omega = \frac{e^{j\omega t}}{2\pi}$$

and $$f(t) = \frac{1}{2\pi} \underline{/\omega t}$$

or $$|f(t)| = \frac{1}{2\pi}$$

Hence $$\frac{1}{2\pi} \rightleftharpoons \delta(\omega)$$

or $$\frac{1}{2} \rightleftharpoons \pi\delta(\omega)$$

where the arrow sign signifies a transform process.

Appendix C: Cooley–Tukey algorithm[11]

This algorithm uses a radix 2 and the number of frequency steps equals the number of time samples N. It expresses the W matrix as a set of γ matrices with $N = 2^\gamma$ where γ is an integer. A signal flow diagram can be drawn to illustrate the computation and is shown in Fig. A.2 for $N = 4$. Here $\gamma = 2$ and it leads to the two square matrices designated as $[F_0]^1$ and $[F_0]^2$.

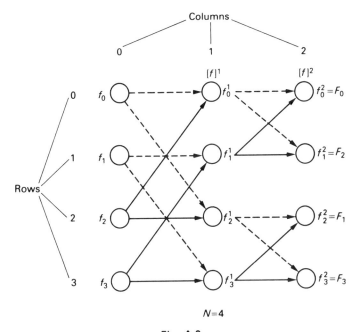

Fig. A.2

The diagram consists of rows and columns with the sample values f_n on the left and the transforms F_n on the right. Intermediate positions are nodal points with numbers in circles representing the powers of the W matrices.

Interconnections consist of dotted lines representing additions and full lines representing multiplications. The diagram is developed using binary arithmetic and certain rules of logic.

The computation of any particular DFT such as F_0 requires the calculation of previous nodal points such as $[F_0]^2$ and $[F_0]^1$. The latter are given by a recursive relation of the form

$$X(c, r) = X(c - 1, r_a) + W^p X(c - 1, r_b)$$

$$\begin{array}{cc} \text{addition} & \text{multiplication} \\ \text{term} & \text{term} \end{array}$$

where X is any node in column c and row r, r_a and r_b are certain row locations, and p is the power to which a W matrix is raised.

Typically, we obtain for the discrete Fourier transform F_0 the expressions

$$F_0 = [f_0]^2 = [f_0]^1 + W^0[f_1]^1$$

$$[f_1]^1 = f_1 + W^0 f_3 = f_1 + f_3 \qquad (W^0 = 1)$$

$$[f_0]^1 = f_0 + W^0 f_2 = f_0 + f_2$$

Hence
$$F_0 = [f_0]^1 + W^0[f_1]^1 = f_0 + f_2 + W^0(f_1 + f_3)$$

or
$$F_0 = f_0 + f_1 + f_2 + f_3$$

and the DFT is given in terms of the four time samples. This result can be checked with the equation given in Chapter 2 for F_m when $m = 0$.

An algorithm for $N = 16$ which employs *decimation-in-time* is shown in Fig. A.3 and it yields the frequency components in regular order. An alternative algorithm which employs *decimation-in-frequency* yields the frequency components in irregular order and will be found elsewhere.[11]

Appendix D: Sampled response

For any non-periodic function $F(t)$, we have

$$F(t) = \frac{1}{2\pi} \int_{-\infty}^{+\infty} F(\omega) e^{j\omega t} d\omega$$

where $F(\omega)$ is a spectral function extending to $\pm \infty$ and is shown in Fig. 5.4.

In order to convert $F(t)$ into a *continuous* function $f(t)$, ω must be restricted to finite values $\pm 2\pi W$ as in Section 5.2. Hence

$$f(t) = \frac{1}{2\pi} \int_{-2\pi W}^{+2\pi W} F(\omega) e^{j\omega t} d\omega$$

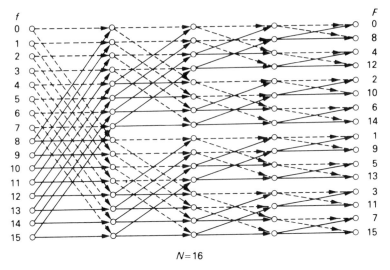

$$N = 16$$

Fig. A.3

Using the result of Section 2.4 for the periodic function $f(t)$, any periodic function $F(\omega)$ is given by

$$F(\omega) = \sum_{-\infty}^{+\infty} C_{-n}\, e^{-j\omega(n/2W)} \quad \text{(by simple analogy)}$$

where $t = 1/2W$ and negative values of n are used for convenience.

But
$$C_{-n} = \frac{1}{4\pi W} \int_{-2\pi W}^{+2\pi W} F(\omega)\, e^{j\omega(n/2W)}\, d\omega \quad \text{(from Section 5.2)}$$

$$= \frac{1}{2W} f(n/2W)$$

Hence
$$F(\omega) = \sum_{-\infty}^{+\infty} \frac{1}{2W} f(n/2W)\, e^{-j\omega(n/2W)}$$

Substituting for $F(\omega)$ in $f(t)$ gives

$$f(t) = \frac{1}{2\pi} \int_{-2\pi W}^{+2\pi W} \left[\sum_{-\infty}^{+\infty} \frac{1}{2W} f(n/2W)\, e^{-j\omega(n/2W)} \right] e^{j\omega t}\, d\omega$$

or
$$f(t) = \frac{1}{4\pi W} \sum_{-\infty}^{+\infty} f(n/2W) \int_{-2\pi W}^{+2\pi W} e^{j\omega(t-n/2W)} d\omega$$

$$= \frac{1}{4\pi W} \sum_{-\infty}^{+\infty} f(n/2W) \left[\frac{e^{j\omega(t-n/2W)}}{j(t-n/2W)} \right]_{-2\pi W}^{+2\pi W}$$

$$= \frac{1}{4\pi W} \sum_{-\infty}^{+\infty} f(n/2W) \left[\frac{e^{j2\pi W(t-n/2W)} - e^{-j2\pi W(t-n/2W)}}{j(t-n/2W)} \right]$$

or
$$f(t) = \sum_{-\infty}^{+\infty} f(n/2W) \frac{\sin x}{x}$$

where $x = 2\pi W(t - n/2W)$.

Comments
1. The peak amplitude of each $(\sin x)/x$ response is $f(n/2W)$ which is also the amplitude at the instant sampled. Hence, each peak response occurs at the instant sampled.
2. The delay of the nth response is $n/2W$ and of the succeeding one $(n+1)/2W$. Hence, the time interval between the peak responses is $1/2W$ which is the sampling period. This confirms the previous result.

Appendix E: Binary arithmetic

In the decimal system, counting is achieved in powers of 10, whereas, in the binary system, powers of 2 are employed. As an example, consider the number 1234.

$$1234 = 1000 + 200 + 30 + 4$$
$$= 1 \times 10^3 + 2 \times 10^2 + 3 \times 10^1 + 4 \times 10^0$$

in the decimal system. This can be arranged as

	10^3	10^2	10^1	10^0
$1234 =$	1	2	3	4

In the binary system, we have

$$1234 = 1024 + 128 + 64 + 16 + 2$$
$$= 1 \times 2^{10} + 0 \times 2^9 + 0 \times 2^8 + 1 \times 2^7 + 1 \times 2^6 + 0 \times 2^5$$
$$+ 1 \times 2^4 + 0 \times 2^3 + 0 \times 2^2 + 1 \times 2^1 + 0 \times 2^0$$

which can be arranged as

	2^{10}	2^9	2^8	2^7	2^6	2^5	2^4	2^3	2^2	2^1	2^0
$1234 =$	1	0	0	1	1	0	1	0	0	1	0

and so the binary number for 1234 is 10011010010.
The various arithmetical processes of addition, subtraction, multiplication, and division are possible in the binary system and are carried out in the same manner as with the decimal system, using the following simple rules.

Addition	*Subtraction*
$0 + 0 = 0$	$0 - 0 = 0$
$0 + 1 = 1$	$1 - 0 = 1$
$1 + 1 = 0$ and carry 1	$1 - 1 = 0$

$1 + 1 + 1$ (carried) $= 1 +$ carry 1

Multiplication	*Division*
$0 \times 0 = 0$	$0 \div 0 = 0$
$0 \times 1 = 0$	$1 \div 1 = 1$
$1 \times 1 = 1$	

Note
The application of binary counting in the field of computers is through the use of logic circuits which perform the various operations of *and, or, not*, etc. More recently, such circuits have been used in connection with electronic switching at telephone exchanges for joining subscribers lines.

Appendix F: Speech production and image processing

Speech production[22,23]
Speech is an analogue signal with a bandwidth from about 50 Hz to around 10 kHz. The vocal organs which produce speech are the lungs, windpipe, larynx, throat, nose, and mouth. During speech production, the vocal tract extending from the throat to the lips is varied in shape by slow movements of the lips, tongue, and jaw which is known as *articulation*.

Speech consists of *voiced* sounds, *unvoiced* sounds, and *plosive* sounds. Voiced sounds which include all the vowels and some of the consonants are due to the vibration of the vocal chords which release 'puffs' of air into the vocal tract. The audio frequencies form a line spectrum and, due to the cavities of the throat, mouth, and nose, resonances occur at about three *formant* frequencies.

Unvoiced sounds such as those for the consonants *f*, *s*, and *p* are produced when air turbulence is caused by a constriction somewhere along the vocal tract. Plosive sounds such as those for the consonants *b* or *d* are caused by stopping the air flow from the lungs by blocking the vocal tract at some point with the tongue or lips.

The human speech system can be modelled as a source of excitation which is followed by a linear filter network. For voiced sounds, the excitation resembles a pulse generator while, for unvoiced sounds, it corresponds to that of a noise generator. The filter network simulates the resonances of the vocal tract which vary with time but may be regarded as quasi-stationary over short time intervals only. It is represented by a suitable transfer function for the network.

Much of the speech waveform contains redundant information and speech processing involves bandwidth-compression techniques with a view to reproducing speech using a much narrower bandwidth than that of the original acoustic spectrum.

Image processing[26]

Electrical images can be processed for transmission, image improvement, or feature extraction. To obtain the link between the optical and electrical processes, consider the translation of an image I_{XY} in the (X, Y) plane into an image i_{xy} in the (x, y) plane. The operation can be represented by

$$i_{xy} = \int_X \int_Y I_{XY}\, h(xX, yY)\, dX\, dY$$

where $h(xX, yY)$ is the point-spread function which determines the spread of light in the image plane due to an infinitely small point source in the (X, Y) plane. If the operator $h(xX, yY)$ is space invariant, the operation can be replaced by the convolution integral

$$i_{xy} = \int_X \int_Y I_{XY}\, h(x - X)(y - Y)\, dX\, dY$$

or

$$i_{xy} = I_{XY} * h(x, y)$$

To obtain the original image, a spatial Fourier transform technique may be obtained by a Fourier transform to the (u, v) plane and obtaining subsequently the inverse Fourier transform to yield I_{XY}. Hence, we obtain

$$I(u, v) = \frac{i(u, v)}{H(u, v)}$$

with

$$I_{XY} = F^{-1}\left[\frac{i(u, v)}{H(u, v)} \right]$$

where F^{-1} represents the inverse Fourier transform and $I(u, v)$, $i(u, v)$, and $H(u, v)$ are the Fourier transforms of I_{XY}, i_{xy}, and $h(x, y)$ respectively.

Appendix G: Walsh functions[38]

These are a set of orthogonal functions defined over the unit interval $(0, 1)$, which is shown in Fig. A.4(a), and take on the values of $+1$ or -1 only. These continuous functions are designated as $wal_j(t)$ where j is the number of zero-crossings in the unit interval $(0, 1)$ on the x-axis which usually represents time t.

The alternative notations of using $cal_s(t)$ for even functions and $sal_s(t)$ for

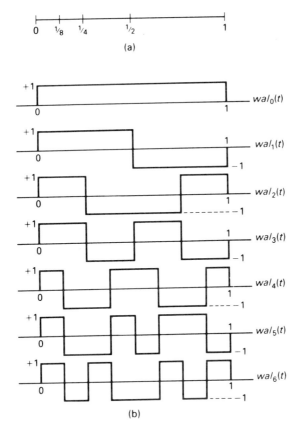

Fig. A.4

odd functions may be employed to designate Walsh functions where s is called the sequency and represents one-half of the average number of zero-crossings per unit interval. Walsh functions and sequency are, broadly speaking, related ideas, as are sinusoidal functions and frequency. The first seven Walsh functions are shown in Fig. A.4(b).

Walsh functions can also be designated by the symmetry properties of the points of division $\frac{1}{2}, \frac{1}{4}, \frac{1}{8}, \ldots, 1/2^n$ in the unit interval shown in Fig. A.4(a). The index j is written as a binary number where 0 and 1 specify even and odd symmetry at a zero-crossing point respectively. The most significant digit is ascribed to the zero-crossing at the $1/2^n$ point of the unit interval and the least significant digit is ascribed to the zero-crossing at the mid-point of the unit interval. For example, $wal_6(t)$ can also be designated as $wal_{110}(t)$ and is illustrated in Fig. A.4(b).

The functions are orthogonal and so we have for two Walsh functions wal_j and wal_k the relationship

$$\int_0^1 wal_j \times wal_k = 0 \qquad j \neq k$$

$$\int_0^1 wal_j \times wal_k = 1 \qquad j = k$$

Furthermore, the product of two Walsh functions $wal_j \times wal_k$ is another Walsh function and is obtained by modulo-2 addition of j and k. Hence, we obtain

$$wal_j \times wal_k = wal_{j \oplus k}$$

where the sign $\oplus$ signifies modulo-2 addition.

It can also be shown that the Walsh functions are related to the Rademacher functions and to the Hadamard matrix. One-dimensional and two-dimensional signal processing by means of the fast Hadamard transform (FHT) and fast Walsh transform (FWT) have drawn great attention in recent years because of the possibility of reducing the bit rate for digital signal transmissions.

In this case, discrete Walsh functions can also be defined by sampling the continuous Walsh functions at N equidistant locations in the unit interval $(0, 1)$ where N is a power of two. A notation used is $wal_j(k)$ where j is the number of sign changes from ∓ 1 to ± 1 and k is the independent variable. Note that j and k can take on N discrete values, e.g. $0, 1, 2, \ldots, N-1$, where $N = 2^n$ and n is an integer.

Appendix H: Discrete source

Information rate

Let the input source to a communication channel (transmitter) be denoted by X and the output source from the communication channel (receiver) be denoted by Y, as in Fig. A.5.

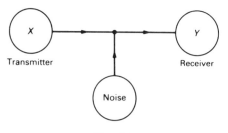

Fig. A.5

For message symbols with probabilities P_i, the entropy of the input source is given by

$$H(X) = -\sum_i P_i \log_2 P_i \tag{1}$$

Similarly, for the output source with symbol probabilities P_j, the entropy is given by

$$H(Y) = -\sum_j P_j \log_2 P_j \tag{2}$$

Now, the conditional probability that the ith symbol was transmitted when the jth symbol was received is $P(i|j)$ and is related to the joint probability $P(i,j)$ where

$$P(i,j) = P_i P(j|i) = P_j P(i|j) \tag{3}$$

Hence, the conditional entropy per symbol received is $H(X|j)$ where

$$H(X|j) = -\sum_i P(i|j) \log_2 P(i|j) \tag{4}$$

from equation (1).

The *average* conditional entropy $H(X|Y)$ for all possible received symbols is given by the sum of $H(X|j)$ over all values of j and weighted with probabilities P_j.

Hence $$H(X|Y) = \sum_j P_j H(X|j) = -\sum_j P_j \sum_i P(i|j) \log_2 P(i|j)$$

or $$H(X|Y) = -\sum_i \sum_j P(i,j) \log_2 P(i|j) \tag{5}$$

from equation (3).

To find the information transmitted by the source, we have

$$\text{information} = \log_2 \left[\frac{a \ posteriori \ \text{probability}}{a \ priori \ \text{probability}} \right]$$

If the ith symbol was transmitted, then the *a priori* probability is P_i and, if the jth symbol was received, the *a posteriori* probability is $P(i|j)$.

Hence $$\text{information} = \log_2 \left[\frac{P(i|j)}{P_i} \right]$$

This is the information transmitted per symbol transmitted and received. To find the information transmitted for all transmitted and received symbols, we must sum over all joint probabilities $P(i,j)$, i.e. $\sum_i \sum_j P(i,j)$.

Hence   information $H = \sum_i \sum_j P(i,j) \log_2 \left[\dfrac{P(i|j)}{P_i} \right]$

or $H = \sum_i \sum_j P(i,j) \log_2 P(i|j) - \sum_i \sum_j P(i,j) \log_2 P_i$

$= \sum_i \sum_j P(i,j) \log_2 P(i|j) - \sum_i P_i \log_2 P_i$

since $\sum_i \sum_j P(i,j) = \sum_i P_i \sum_j P(j|i) = \sum_i P_i$ from equation (3).

Hence $H = -H(X|Y) + H(X) = H(X) - H(X|Y)$

from equations (1) and (5).

If entropy rates are used instead of entropies, it yields the transmission rate R where

$$R = H' = H'(X) - H'(X|Y)$$

If the discrete source transmits n symbols where x_i is the ith symbol transmitted and y_j is the jth symbol received, as illustrated in Fig. A.6, the

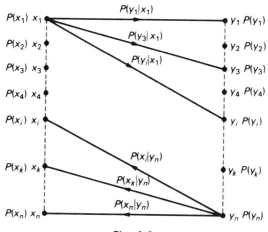

Fig. A.6

previous results can be summarised as follows

$$H(X) = - \sum_{i=1}^{n} P(x_i) \log_2 P(x_i)$$

$$H(Y) = - \sum_{j=1}^{n} P(y_j) \log_2 P(y_j)$$

$$H(X|Y) = - \sum_{i=1}^{n} \sum_{j=1}^{n} P(x_i, y_j) \log_2 P(x_i|y_j)$$

$$H(Y|X) = - \sum_{i=1}^{n} \sum_{j=1}^{n} P(x_i, y_j) \log_2 P(y_j|x_i)$$

$$I(X; Y) = H(X) - H(X|Y) = H(Y) - H(Y|X)$$

where $P(x_i, y_j) = P(x_i)P(y_j|x_i) = P(y_j)P(x_i|y_j)$ is the joint probability and $I(X; Y)$ is known as the *mutual information*.

Appendix I: Continuous source

Information rate
Analogous to the definition of entropy for a discrete source, the entropy of a continuous random variable x can be defined by

$$H(X) = \int_{-\infty}^{+\infty} P(x) \log_2 P(x) \, dx$$

and it can be shown that, if the variance of x is σ^2, the probability density function $P(x)$, which gives the maximum entropy, is Gaussian and of the form

$$P(x) = \frac{1}{\sigma\sqrt{2\pi}} e^{-x^2/2\sigma^2}$$

with
$$H(X) = \log_2 \sqrt{2\pi e \sigma^2} \text{ bits}$$

For band-limited white noise, it can be specified by $2W$ samples per second if W is the maximum bandwidth according to the sampling theorem. Also, as the samples are uncorrelated, the entropy rate becomes

$$H'(X) = 2WH(X) = W \log_2(2\pi e \sigma^2)$$

and, for an average noise power $N = \sigma^2$, we obtain

$$H'(X) = W \log_2 2\pi e N \text{ bits/s}$$

Channel capacity
Analogous to the discrete case, this is defined as

$$C = \text{maximum rate of } [H(X) - H(X|Y)]$$

with
$$H(X) = - \int_{-\infty}^{+\infty} P(x) \log_2 P(x) \, dx$$

$$H(X|Y) = - \int_{-\infty}^{+\infty} \int_{-\infty}^{+\infty} P(x, y) \log_2 P(x|y) \, dx \, dy$$

Hence $C = $ maximum rate of $\left[-\displaystyle\int_{-\infty}^{+\infty} P(x) \log_2 P(x) \, dx \right.$

$$\left. + \int_{-\infty}^{+\infty} \int_{-\infty}^{+\infty} P(x,y) \log_2 P(x|y) \, dx \, dy \right]$$

or $C = $ maximum rate of $\left[\displaystyle\int_{-\infty}^{+\infty} \int_{-\infty}^{+\infty} P(x,y) \frac{\log_2 P(x,y)}{P(x) P(y)} \, dx \, dy \right]$

which is *symmetrical* in x and y and so we also have

$$C = \text{maximum rate of } [H(Y) - H(Y|X)]$$

If the output $y(t)$ is linearly related to the input $x(t)$ by $y(t) = x(t) + n(t)$ where $n(t)$ is the noise which is statistically independent of $x(t)$, i.e. the noise is additive to the signal, it can be shown that $H(Y|X)$ is equal to $H(N)$ which is the entropy of the noise. Hence, C is made maximal by maximising $H(Y)$ relative to the input channel and we obtain $C = $ maximum rate of $[H(Y) - H(N)]$.

For a channel band-limited to W Hz with a mean-square noise power of N and a mean-square signal power of S, $H(Y)$ will be a maximum if $y(t)$ is Gaussian with a mean-square value of $\sigma^2 = (S + N)$. Hence, we obtain

$$H(N) = W \log_2 (2\pi e N)$$

$$[H(Y)]_{\text{max}} = W \log_2 [2\pi e (S + N)]$$

with $C = W \log_2 [2\pi e (S + N)] - W \log_2 (2\pi e N)$

or $C = W \log_2 \left(\dfrac{S + N}{N} \right)$ bits/s

This result shows that, by using signals coded to have the properties of Gaussian noise, it is possible to achieve the communication capacity C, or slightly less, with arbitrarily small error.

Rate-distortion function[30,39,40]

A continuous source which can assume an infinite set of values would require a channel with an infinite capacity for the true transmission of its information. In practice, an *exact* transmission is not required but only an acceptable transmission in accordance with some defined *fidelity criterion D*.

For a transmitted message $x(t)$ and a received message $y(t)$, the requirement is for the *minimum* rate with which this can be achieved, subject to the required fidelity criterion which is denoted by $d(x,y)$ since it depends on the joint probability density function of the source variable $x(t)$ and the required message $y(t)$.

The minimum rate for generating information subject to this fidelity criterion is given by

$$R(D) = \min_{P(y|x)} \int_{-\infty}^{+\infty} \int_{-\infty}^{+\infty} P(x, y) \log_2 \frac{P(x, y)}{P(x)P(y)} \, dx \, dy$$

$$D = \int_{-\infty}^{+\infty} \int_{-\infty}^{+\infty} d(x, y) P(x, y) \, dx \, dy$$

where $R(D)$ is defined as the *rate-distortion* function and is evaluated over all possible values of the conditional probability density function $P(y|x)$ which determines the likelihood of a given received message $y(t)$, given that the message $x(t)$ was transmitted.

A fidelity criterion which may be employed is the mean-squared error criterion $d(x, y) = [x(t) - y(t)]^2$. In other cases, such as speech or television, a frequency weighted error criterion is more suitable, while in digital systems, the bit error rate is more suitable or meaningful.

Shannon has shown that, for a Gaussian white noise source of average message power Q and source bandwidth W_s, we obtain

$$R(D) = W_s \log_2 (Q/D)$$

where D is the mean-squared error or fidelity criterion used. Furthermore, a necessary condition for transmitting a message $x(t)$ over a noisy channel with capacity C and mean-squared error criterion D is given by $R(D) \leqslant C$.

The importance of the rate-distortion function is due to the fact that it gives the *minimum* channel capacity required for transmission with the given fidelity criterion. Hence, it is useful for comparing both source coding techniques and channel coding techniques. In the former case, it has been shown that, for bit rates greater than a certain minimum $R_{\min}$, the rate-distortion function is given by

$$R(D) = \tfrac{1}{2} \log_2 (\sigma_m^2 / D)$$

where $0 < D < \sigma_m^2$ and σ_m^2 is the minimum mean-squared prediction error one step ahead.

In the case of channel coding, it has been shown that, for a band-limited Gaussian source with a mean-squared error criterion, we have

$$\frac{Q}{D} = \left(1 + \frac{W_s}{W_c} \frac{S}{N_0 W_s} \right)^{W_c/W_s}$$

where W_c is the channel bandwidth, W_s is the source bandwidth, S is the signal power, and N_0 is the noise power spectral density. The equation can be used to compare various modulation techniques, in terms of the bandwidth expansion ratio (W_c/W_s), for a given signal-to-noise ratio in the channel.

Index